The Sensorium of the Drone and Communities

The Sensorium of the Drone and Communities

Kathrin Maurer

The MIT Press
Cambridge, Massachusetts
London, England

The MIT Press would like to thank the anonymous peer reviewers who provided comments on drafts of this book. The generous work of academic experts is essential for establishing the authority and quality of our publications. We acknowledge with gratitude the contributions of these otherwise uncredited readers.

This book was set in Stone Serif and Stone Sans by Westchester Publishing Services. Printed and bound in the United States of America.

Library of Congress Cataloging-in-Publication Data

Names: Maurer, Kathrin, author.
Title: The sensorium of the drone and communities / Kathrin Maurer.
Description: Cambridge, Massachusetts : The MIT Press, [2023] | Includes
    bibliographical references and index.
Identifiers: LCCN 2022048658 (print) | LCCN 2022048659 (ebook) |
    ISBN 9780262545907 (paperback) | ISBN 9780262374897 (epub) |
    ISBN 9780262374903 (pdf)
Subjects: LCSH: Drone aircraft—Psychological aspects. | Drone aircraft—
    Societies, etc. | Sociotechnical systems. | Gemeinschaft and Gesellschaft
    (Sociology) | Sensor networks. | Electronic surveillance.
Classification: LCC TL685.35 .M38 2023 (print) | LCC TL685.35 (ebook) |
    DDC 629.133/39—dc23/eng/20230222
LC record available at https://lccn.loc.gov/2022048658
LC ebook record available at https://lccn.loc.gov/2022048659

10 9 8 7 6 5 4 3 2 1

For Christian

Contents

Contents

Acknowledgments

This book revolves around the idea that drones are sensorial assemblages entwining humans and nonhumans. While writing this book, I gradually learned that anything can be an assemblage, and this book became one too. Not only because it passed through many amorphous stages before it went to press but also because it took innumerable interactions, relations, networks, entanglements, and interfaces to write it. I am deeply indebted to this book-assemblage and its generative processes. As I experiment with some of the philosophical tenets of new materialism and posthumanism in this book, let me begin by thanking all the things with nonhuman agency that played an important role in the book-assemblage (although I will have to be very selective).

I am deeply indebted to the different surfaces on which this book was written. The wooden kitchen table in our New York City apartment. The smooth surface of my Danish Design desk, which stayed cool despite world-shattering events. The library reading rooms that never failed to trigger feelings of calm and serenity as I enjoyed the privilege of sitting in spaces such as Columbia University's Butler Library and Avery Hall, New York's Bobst Library, the New York Public Library, and the Royal Library in Copenhagen. Thanks to all the art books I could read and touch in those libraries (especially during the time before the pandemic). Thanks also to my computer for not losing my stuff, and for never, ever criticizing me. I am grateful to all my pens and pencils, since I am a person that still drafts chapters on paper. And to all the different smells that are connected to this book: the smell of the A train I took in the early mornings from 181st Street to New York University during my research stay in 2019–2020; the smell of green summers in Denmark, which persisted even though the world was in pandemic lockdown; the smell of coffee (without you, this book would not have been possible). On that note, I am grateful to all the cafés that sheltered me while they were still open: Hungarian Pastry Shop, Uptown Garrison, Buuni Café, Black Diamond Café, and Krøyers Café.

But much more important than these things were the people that helped me bring this book to life. I am absolutely indebted to my editors at the MIT Press, Lillian Dunaj, Noah J. Springer, and Doug Sery, who believed in this book and guided me through it with patience and support. I also thank the peer reviewers, whose thoroughness and academic excellence truly helped me to improve and sharpen my ideas. I am grateful for the excellent editorial work by Merl Fluin (formerly Storr): her superb editing skills were extremely inspirational and beneficial for my writing process. Thanks also to all the artists that were kind enough to give me permission to use their wonderful artworks in this book.

Further, I am grateful to the colleagues affiliated with my research cluster Drone Imaginaries and Communities, sponsored by the Independent Research Fund Denmark: Lila Lee-Morrison, Rikke Munck Petersen, Dominique Routhier, Kristin Veel, and Kassandra Charlotte Wellendorf. Their interest in anything that has to do with the drone and their comments on draft chapters were indispensable. The many drone aficionados in my research network Drones and Aesthetics, sponsored by the Independent Research Fund Denmark, have been truly inspirational too. To name but a few, Ignacio Acosta, Daniela Agostinho, Christian Ulrik Andersen, Emanuele Nicolò Andreoli, Svea Braeunert, Steen Ledet Christiansen, Rasmus Degnbol, Andreas Immanuel Graae, Dan Gettinger, Mareile Kaufmann, Ina Neddermeyer, Lotte Philipsen, Søren Bro Pold, Thomas Stubblefield, Tomas van Houtryve, and Louise Wolthers. I would also like to thank my colleagues at the University of Southern Denmark's Center for Culture and Technology, whose collaboration I greatly appreciate: Dylan Cawthorne, Stig Børsen Hansen, Casper Sylvest, and Bo Kampmann Walter.

This book is partly a product of my metamorphosis from a German studies scholar to a scholar with an interest in technology. I thank my colleagues and my students at the University of Southern Denmark, at both the Institute for the Study of Culture and the Danish Institute for Advanced Studies, for supporting my academic transformation. I would also like to thank my hosts at New York University—Elisabeth Strowick and the Digital Theory Lab, and Leif Weatherby—for welcoming me during my sabbatical, and the Carlsberg Foundation for its financial support for that sabbatical. I am grateful to have met Nancy Berlinger at The Hastings Center, who became not only a research colleague but also a close friend. I thank Sophie Wennerscheid and Elizabeth Wolff for letting me work in their apartments when mine became uninhabitable due to construction noise.

But there were many other people that helped this book come into being. I cannot thank them all individually, but I would like to mention some of them and thank them for their academic inspiration and friendship: Michele Barker, Mercedes Bunz, Rosi Braidotti, Paul Cureton, David Howes, Maximilian Jablonowski, Ole B. Jensen, Christine Kanz, Sofie Kluge, Charlotte Kroløkke, Claudette Lauzon, Joseph Lemelin, Karen Hvidtfeldt, Matthew Miller, Anna Munster, Peter Schwartz, Ulrik Pagh Schultz, Nanna Bonde Thylstrup, Jutta Weber, and Joanna Zylinska.

Finally, I am indebted to my family. To our two teenagers, Siri and Leo, for their great curiosity and all the laughs we are able to share together. And to my husband, Christian Rostbøll. Christian, your interest in my projects has remained undimmed throughout all our years of marriage, and I am grateful for all our academic discussions. But I am also thankful that I can share the experience of more profane things with you, the failures, occasional successes, jokes, joys, disappointments, worries—in other words, all the stuff that life throws at me. It is to you that I dedicate this book.

Introduction

Drones can be blobs. In Agi Haines's art project *Drones with Desires* (2015), the drone is bubblelike, globular, and slimy looking. A large pink balloon that resembles a human organ, with dark red and bluish blotches, hovers in the gallery.[1] The artwork does not portray the cold, smooth drone one would associate with precision remote-targeting technology. Instead, this drone is amorphous, fleshy, and alive. To create this drone-blob, Haines scanned her own brain and translated its visual anatomy into artificial neural networks that control the drone's motions through sensory inputs. We might understand Haines's drone-blob as referring to the diffuse, dehumanizing, and monstrous violence of military drones. However, I do not read this blob as epitomizing the "techno-bestiary" of the drone.[2] For me, its "blobbiness" exemplifies the drone's potential for antiscopic and multisensorial sensing. The blob and its datafied, tactile, and kinetic modes of sensing point to this book's main idea: drones are much more than aerial cyclopes with pointed vertical vision; they embody synesthetic sensoria with a plethora of heterogeneous sensing modes.

Drone technology has garnered critical attention across disciplinary fields, from engineering to the social sciences and humanities. The first wave of drone scholarship was pivotal in initiating the debate on the emergence and routinization of this technology in the military. In doing so, it privileged the idea of the "scopic regime" in its analyses of the connection between vision and power. These early researchers defined the drone's scopic regime as a militarized system of hypervisuality that effected a vertical, hierarchical, and totalized power relation between the drone operator and the surveilled target. But while the scopic regime is certainly a key characteristic of drone vision, this book broadens the drone's spectrum of perception by drawing on the notion of the sensorium. Taking an

Figure 0.1
Agi Haines, *Drones with Desires* (2015). © Agi Haines.

aesthetic approach to the analysis of civilian drones, the book understands
the sensorium of the drone as a complex, multimedia, synesthetic sensing
assemblage, in which the human agent is enmeshed with the technical
apparatus.[3] Drone sensoria sense in many more ways than the notion of
the scopic regime suggests. Drone sensing can be embodied, datafied, flattened, volumetric, or swarmlike, and these different modes of sensing often
connect to other more-than-optical sensual registers, such as sound, touch,
smell, temperature, and movement. This sensorium disrupts the idea of the
strawlike, surgical visual regime of the drone as an instrument of precision
and pointed invasion.[4]

I am specifically interested in how artists experiment with these drone sensoria in their artworks. Many contemporary artists use the civilian drone and
its sensorial potentials as an aesthetic and creative medium. There are literary novels that engage with the sensorium of the drone swarm, and there
is visual art that operates with forms of embodied drone sensing. But it is
important to note that these aesthetic drone sensoria also go beyond the

visceral level of machinic sensing. It is imperative to my book's agenda to understand that these drone sensoria also have a collective dimension: they are about communities.

Drone researchers to date have mainly discussed communities in a military context.[5] As products of the visual politics of the scopic regime, such communities feature the clear oppositions that the regime entails: top/bottom, target/nontarget, observer/nonobserver, visible/invisible, machine/human.[6] Unlike this previous research, however, I argue that the aesthetic sensoria of civilian drones can construct communities on the basis of decentralized, network-like, and fluid sensing processes.[7] In this context, *constructing* means that in aesthetic works the sensorial assemblage of the drone can create imaginaries of communities. This notion of the imaginary connects in some ways with the work of Charles Taylor, who points to the idea that cultural practices shape social imaginaries.[8] However, I use the term *imaginary* slightly differently. Taylor conducts a macroanalysis of how Western cultural forms and moral normative discourses display the social imaginaries of modernity. I am more interested in the fluid, dynamic, and amorphous imaginaries of communities that are generated by the aesthetic sensorium of the drone in a plethora of different artworks. In addition to these aesthetic imaginaries of communities, I also focus on how civilian drone sensoria can generate "real" communities. There are real-world communities where the drone plays a constitutive role, such as hobby drone communities, social movements, and groups of political activists.[9] But these communities too—and here I draw on Benedict Anderson—have an imaginary dimension, as they revolve around ideas and narratives about the drone.[10] As I discuss in chapter 1, which sets out my theoretical framework and develops my understanding of the notion of community, I am especially interested in imaginaries of communities that highlight machine–human assemblages, swarms, multitudes, and data-calculable publics. The drone sensorium can engender visions of communities that suggest new forms of bonding between machinic and human environments. It can create imaginaries of posthuman collectives that are no longer exclusively defined by human subjectivity and identity but rather are manifested in looser constellations between the human and the nonhuman—the zoological, the earth, and the machine.

The Sensorium of the Drone and Communities thus explores the sensorium of civilian, commercial, and amateur drones, although I remain highly aware that military and civilian remote sensing technologies are deeply intertwined. In comparison with military drones, civilian drones are still underresearched, particularly in the fields of humanities, media studies, and aesthetics.[11] Cultural approaches to civilian drones often highlight their close relationship with the military, considering domestic drones in light of a "boomerang effect"[12] from military drone applications. Domestic drones thus supposedly exemplify media theorist Friedrich Kittler's contention that the entertainment industry's media products inherently belong to the military-industrial complex.[13] I acknowledge these connections, but I try to work beyond them by looking at the drone not only as harmful but also as playful. Civilian drones are not just about the militarization of the everyday; they have their own creative power. One goal of this book is to (partially) free the drone from its military stigma, and to detect its creative, life-affirming potential without fetishizing it or forgetting its military baggage. Recent scholars have discussed civilian drones beyond the military genealogical fallacy, and this book is greatly inspired by their work.[14] This does not mean that I am dismissive of the military origins of drone technology but rather that I am attempting to interpret drone sensing and its shaping of communities without exclusively highlighting its military side. One can always trace military contexts and practices when one is dealing with drones. For me, it is a matter of observation: this book attempts to highlight the creative, constructive, and affirmative aspects of the drone, although I always keep the drone's potentially violent practices in mind. The goal is to read drone technology in terms of the creation of a unique sensorium of which the visual is only a part, and to consider drones' capacity to establish communities that interrupt and contradict the power relations of drone violence.[15]

My book stretches the idea of what a drone is. Technically, a drone is defined as a flying object—an unmanned aerial vehicle that moves by remote control—or alternatively, as an unmanned ground or underwater vehicle, as drones can also crawl, swim, and dive. But for me as a scholar of visual culture, art, and literature, a drone is not just a technical apparatus or instrument, it is about discourse. Drones do not operate somewhere "out there," independently of us; they are our sensing avatars, deeply enmeshed with cultural and political contexts. As technocultural assemblages, they are attached to our affectivities, to their surrounding materialities, and to their embodiments. Ole Jensen has suggested that drones can act as "epistemological engines" that provide knowledge about our world.[16]

Aisthêsis, Art as Sensorium, and Aesthetics as Critical Discourse

To understand these worldmaking powers of the civilian drone, this book takes an aesthetic approach to technology, working with two intersecting understandings of aesthetics, namely, aisthêsis and aesthetics as a critical discourse. Following the original meaning of the Greek word, aisthêsis refers to aesthetics as a way of experiencing, as Alexander Gottlieb von Baumgarten proposed in *Aesthetica*.[17] For Baumgarten, the sensuous and aesthetic realm constitutes a form of knowledge that has an epistemological power of its own, one that is based on sensing and experiencing art. Notably, Baumgarten suggests that one can train this form of knowledge via technology.[18] His idea of aesthetic training involves affective experiences with the "weapons of the senses."[19] These are instruments of visual, thermal, and sonic sensing, such as "magnifying glasses, binoculars, artificial ears, barometer, thermometer."[20] For Baumgarten, aisthêsis thus entails a form of sensory perception that can be drilled, enhanced, and fostered by technology. This conception of aisthêsis, which grasps the experientiality of technology, is vital for my work because I explore the sensual and affective dimension of the drone and the material–bodily entanglements between the drone and its agent. My focus on aisthêsis implies an unearthing of the ways in which drones sense—for example, what sensory signals, images, and data they produce, and what sensorial affectivities drones can trigger in the human agent. But aisthêsis does not only describe the technosensuality of a technological object. Drone artists also experiment with the aisthêsis of the drone by turning their artworks into sensoria, thereby making drone-aisthêsis a quintessential aspect of their art.

Digital artists have had an interest in the topic of machinic aisthêsis for quite some time. During the 1980s and 1990s, many began to explore the relationship between bodily experiences and electronics,[21] and "new forms of subjectivity were theorized, from cyborgs to digital *flâneurs* to networked hivemind."[22] Often these artistic sensoria were synesthetic, speaking to visual, auditory, olfactory, embodied, and tactile sensing registers.[23] Today, many artists are engaging with technological sensoria by critically intervening in the "techno sensual comfortzone,"[24] investigating the commoditization of technosensuality as a part of our everyday lives (for example, in the iPhone, biocomputing, and touch screens).[25] Civilian drones do not really form a part of this technological comfort zone: they do not emit a sense of ease. Indeed, domestic drones often emanate an aura of the uncanny, creepy, or threatening. But this makes drone artists' negotiations with the drone sensorium all the more interesting. Drone artists rarely stop at the affective level when dealing with the drone sensorium; many of them do much more than just present the drone as a medium with "fun" sensorial possibilities. Rather, they engage with the discursivities of sensing. Drone artists often perform negotiations with, observations of, and interventions in this technology, thereby articulating the second dimension of aesthetics that is at stake in this book: aesthetics as a critical discourse.

Aesthetics epitomizes autonomous representation. It does not have to fulfill any sense-making rules. It is precisely this freedom to observe the world beyond instrumental codes that can render aesthetics a powerful discourse of critique. As modern theories of aesthetics often emphasize, art can (but does not have to) be a medium through which to voice critique. From the German Romantics to the Frankfurt School to systems theory, aesthetics offers the conditions in which to observe the world differently, that is, noninstrumentally, affectively, and nondiscursively. This power of aesthetics is crucial for my interpretation of drone art. Aesthetic negotiations can reflect the vulnerability, uncertainty, and fallibility of drone technology and can obfuscate techno-optimistic narratives about the drone's precision.[26] In drone artists' work, drone sensoria can evoke dissensus: they can be negotiated, criticized, ironized, queered, celebrated, and made strange.[27] Of course, we have to be aware that this view of aesthetics—as critical—also undermines (at least momentarily) its autonomy, since it must attain a heteronomous determination in order to create dissensus.

This book discusses drone artworks in very different aesthetic genres, including literature, film, photography, and visual art installations and performances. In science fiction novels, literary diaries, dramas, and contemporary prose, for example, drones can occur as protagonists, violent machines, or technical gadgets. Often these literary works highlight drones' military connections. Atef Abu Saif's *The Drone Eats with Me: Diaries from a City under Fire* (2015) narrates the experience of the everyday threat of drone bombardment; the one-woman drama *Grounded* (2013), by the British playwright George Brant, consists of a monologue by a former F-16 fighter pilot who became a military drone sensor operator.[28] In popular culture too, the military drone has played starring roles, such as in Gavin Hood's movie *Eye in the Sky* (2015). In the realm of experimental visual art, Omer Fast's

5000 Feet Is the Best (2011), which deals with the traumatic experiences of drone pilots, has garnered much attention.

But instead of art about the military drone, I explore art with and about domestic consumer drones. Literary works, visual installations, films, artistic products from the hobby drone scene, and artworks by political activists can all serve as a repository of examples that demonstrate the multisensoriality of the drone and shed light on its community-shaping powers. For example, there have been literary novels about civilian drones, such as Tom Hillenbrand's sci-fi *Drone State* (*Drohnenland*, 2015), in which drones function as telepresence devices and provide embodied forms of communication.[29] There is the drone surveillance installation *Hansel and Gretel* by Jacques Herzog, Pierre de Meuron, and Ai Weiwei, staged at the Armory in New York City in 2017, in which a drone tracked and observed its visitors. There have also been drone ballets, such as *Networks of Tomorrow* (2021), and drone operas, such as Matthew Sleeth's *A Drone Opera* (2019). While I concentrate on contemporary art on and with domestic drones, I also look back into history.[30] An excursion into nineteenth-century hot-air balloons and their scholarly poetic descriptions shows that they can be seen as early drones. A discussion of Ernst Jünger's novel *The Glass Bees* (*Die gläsernen Bienen*, 1957) provides me with another trajectory to historicize drones.[31] This historicization is necessary if we are to rethink and intervene in narratives that celebrate drone technology as a game-changing, state-of-the-art technology.

Although I frequently refer to *drone art* throughout this book, I am hesitant to establish it as an aesthetic genre; I use the expression more as a working term. The artworks and artists that engage with drones in aesthetic contexts constitute a heterogeneous group that includes political activists, established photographers, amateurs, pop-culture artists, literary authors, and avant-garde filmmakers. Even in the context of military drones, which have been researched more systematically,[32] I find it difficult to speak of drone art as a genre, as these artists also come from diverse backgrounds, work with different media, and relate different narratives about the drone.[33] In the field of civilian drones, the art scene is even more dispersed, since these artists are often very close to hobbyist and do-it-yourself drone communities. Indeed, this amateurship is a decisive factor, and I discuss a variety of drone artworks that have emerged from the hobby drone scene (such as dronies and drone films about the pandemic posted on YouTube). The drone as an aesthetic medium is fairly accessible to amateurs—commercial drones have become affordable and attainable for many people—and drones' usage in film, photography, and journalism can have a "democratizing" effect insofar as aerial shots no longer have to be made with expensive equipment, such as helicopters and airplanes.[34]

This amateurship by no means implies an inferior aesthetic quality compared with "classic" artworks about drones, such as visual works by Trevor Paglen and James Bridle. But the quality of the "poor image,"[35] as Ariella Azoulay has termed it, does allow me to think about the sensorium of the civilian drone outside the military box. Often, the drone imaginaries of amateur artists who engage with civilian drones articulate a critique of the "technological sublime"[36]—for example, the aestheticizing view of technological inventions as grandiose and monumental. In particular, hobby drone artists undermine the ostentation and fetishization of the drone by producing "quirky" aesthetic images.[37] This book opens up—albeit not exclusively—to "B-grade" aesthetics, amateur art, low-budget productions, and mass-culture phenomena. In this regard, my theoretical muse has been Walter Benjamin and his dialectical embrace of mass culture and technology.

The Dialectics of the Drone Sensorium

For Benjamin, mass media shape new creative processes and alter our ways of perception, reception, and interpretation. Photography, for example, evokes the destruction of the "aura" and demands an aesthetic sensing that is no longer based on contemplation, ritual, or cult. Rather, photographs trigger distraction, thereby undoing the traditional aesthetic categories of contemplation, uniqueness, originality, and myth. Benjamin welcomes these new media technologies and considers their destructive force as the condition of possibility for construction, that is, for a utopian perspective on a different society. Depending on how one interprets Benjamin's philosophemes about media technology, this constructive moment of technology can be incorporated into a Marxist theoretical frame, a messianic theological horizon, or a theory of aesthetic experience (or a combination of all three). For me, it is not so important to determine which type of utopia is at stake in Benjamin's thinking. Rather, what is key is that his philosophy marks the advent of new technologies as a productive and empowering moment, and that art in league with these technologies can construct, shape, and constitute imaginaries of communities.

Drone art can offer new perspectives on communities and human–machine relationships, as it can engender utopian imaginaries of social bonding in human–machine collectives. As the chapters of this book will show, these imaginaries can take the shape of nonessentialist, nonhuman-centered, and nonbinary communities. But it is important to recognize that technology's utopian potential has a flipside. For Benjamin, this utopian moment was always doomed beneath the shadow of fascism, and he noted that during his time, society was not sufficiently mature to deal with technology.[38] Benjamin saw war as the epitome of this condition of human immaturity, since war demonstrates the perversion and destructiveness of technology. As long as society is ruled by fascism and capitalism, it will remain too immature to use technology humanely and organically.

My investigations of various drone artworks in the following chapters engage with the Janus face of drone technology. Drones and their aesthetic sensoria and imaginaries are masters of dialectics. Like reversible images that can change from a duck into a rabbit or from an old woman into a young woman, drones can flip between life and death in the twinkling of an eye.[39] In this sense, the military drone and the civilian drone are not understood as strict opposites but rather as two sides of the same coin. For example, as I will show in chapter 5, there are drones that are used for farming and agriculture. Although they have a vital and life-affirming agenda, I discuss how their operations (detecting bad and good crops, exterminating parasites) are informed by military optics. Or take the drones that monitored crowds during the pandemic (chapter 7): their surveillance was for the common good, but it could easily constitute a breach of privacy and was ingrained with "everyday militarism."[40] It is imperative for me not to turn a blind eye to the militarism of the drone and its capacity to construct inhumane imaginaries of communities. Nevertheless, this book aims to highlight the affirmative, utopian, and creative sides of drone technology. As Donna Haraway notes of cyborgs, "The main trouble with cyborgs, of course, is that they are the illegitimate offspring of militarism and patriarchal capitalism, not to mention state socialism. But illegitimate offspring are often exceedingly unfaithful to their origins. Their fathers, after all, are inessential."[41] Haraway's words reflect the spirit of this book. Yes, drones as cyborgs are the products of capitalist technology companies and the military. But they can also be unfaithful to their origins and develop their own contexts, frameworks, and uses that resist those of their creators. Drone sensoria are epistemic prisms that reveal how drones as machinic–human assemblages construct, have an impact on, and shape the worlds and communities we live in.

Chapter Outline

In the theoretical and historical chapter 1 in Part I, I expand on key terms, such as *drone sensorium, assemblage, aesthetics*, and *communities*. All of the subsequent chapters focus on specific sensing modes of the drone sensorium within an aesthetic context. Each chapter traces a mode of nonscopic drone sensing—embodied, facial-datafied, flattened, volumetric, swarmlike, and viral sensing—and their respective imaginaries of communities. These explorations of the various nonscopic modes of drone sensing are organized against three larger discursive frames, namely, the body (Part II), the earth (Part III), and the nonhuman (Part IV). Although these frames, and their respective sensorial registers, certainly overlap, they function here as hermeneutic backdrops to orient us amid the variety of aesthetic drone sensoria and their imaginaries of communities.

Part II investigates the relationships between the drone, the body, and communities. For Michel Foucault, the body is more than a biological-physiological organism: "[The] body is directly involved in a political field; power relations have an immediate hold upon it; they invest it, mark it, train it, torture it, force it to carry out tasks, to perform ceremonies, to emit signs."[42] The body is always embedded in discourses of power, conforming to them, negotiating with them, and resisting them.[43] This is also true for my analysis of drone-body sensoria, and I show that these technobodies must be seen as expressions of political and cultural discourses.

Chapter 2 takes the drone's embodied sensing as a key sensory concept. In contrast to drones' frequent association with distance, remoteness, and disembodiment, I argue that sensing with amateur drones can be close, intimate, and immersed. By discussing dronies, drone races, and drone art installations, I show that drones can perform embodied sensing beyond the scopic regime of vision. Embodied sensing cannot be reduced to one sense only: it affects the whole body in a synesthetic experience involving vision, touch, sound, and kinetic sensing.

Chapter 3 expands this discussion of the role of the body within the drone sensorium, shifting attention to the field of facial recognition. How do civilian drones with facial recognition software sense, read, and recognize faces?

How do artists react to these technological developments? The drone sensorium with facial recognition can no longer be grasped as a vertical mode of scopic perception. Rather, fictive imaginaries suggest that the drone and its facial sensing are networked, datafied, and multidirectional. Here I discuss artworks that portray the drone's capacity to see faces as a special form of social bonding, in which the drone and the human forge an affective alliance. Other artists—by far the majority—warn us about drones with facial recognition, dismantling their failures, biases, and violence.

Part III focuses on drone sensoria that sense the earth—what lies on, underneath, and above its surface. The drone as a remote sensing technology can provide data about climate change, geology, soil quality, and atmospheric conditions. These earth drones are used by scientists, farmers, zoologists, and environmentalists. But artists also engage with earth drones, working with the drone as an eco-medium that generates knowledge about the condition of the earth and enmeshes us with the planet. For example, in Kim Stanley Robinson's cli-fi novel *The Ministry for the Future* (2020), drones are a medium of ecoterrorism to protect the earth from climate collapse.[44] In addition, there are many experimental visual artists that explore drones as tools to sense, monitor, and observe the earth. In these works, the earth is not a distant place external to us. Rather, it represents what Yuk Hui calls a "milieu"[45] closely interconnected with technology. Hui's critique of the dualism between ecology (nature, the organic) and machines (the mechanistic) is guiding here, as his philosophy promotes an understanding of these as interconnected milieus, spheres, and environments.[46] The chapters in this section show that the drone brings us closer to the earth and makes us reflect critically on human-centered views of the planet.

Chapter 4 provides a new take on the flattening drone view of the earth. Traditionally, the drone's flattened view (and aerial perspectives in general) has been connected to a violent gaze based on abstraction, gridlike cartography, and dehumanization. However, this chapter connects the drone to the romantic history of early hot-air and gas ballooning, and it shows that in contemporary visual artworks the drone's flattened sensing can envision planetary communities. In these imaginaries of communities, humans appropriate the earth not as a specific territory that belongs only to some, but as a common planetary space that needs to be protected.

Chapter 5 traces the earth drone by highlighting another sensorial register: the volumetric, three-dimensional, and atmospheric. These earth drones are interfaces and data processors that can look up from the ground to the sky, monitor the earthy surface (soil) and its in-between spaces, and sense beneath its surface and into the ground. Accordingly, drones can generate sensoria that suggest the three-dimensionality of space, moving away from the scopic (vertical) and flattening cartographic paradigm.[47] Analyzing works of drone art, I show that this form of sensing enables imaginaries of communities that renegotiate our anthropocentric relationship with the earth.

Part IV approaches the drone sensorium from the perspective of nonhuman and zoological modes of sensing and shaping communities. The military and its engineers have a knack for naming drones after swarm animals (for example, Gnat, Killer-bee, or Airspeed Queen Wasp).[48] In fact, the relation between animals and warfare is an interesting one. Often, the animal world serves as a trope to describe the cruelty of war and dehumanize the enemy.[49] But drones have been linked to swarms in the domestic sphere too, and artists have worked with domestic drones, insects, and swarm motifs. Consider, for example, Björn Schülke's *Spider Drone* (2011), a moving, spiderlike, remote-controlled surveillance sculpture, or Roman Signer's analogue experiments with a swarm of small, battery-powered helicopters in *56 Small Helicopters* (*56 Kleine Helikopter*, 2011).

Chapter 6 investigates civilian drones' swarm sensing in popular film, German modernist prose, and protest movements. The drone swarm in "Hated in the Nation," a 2016 episode of the Netflix series *Black Mirror*, stages the swarm as the horrible and bestial "other." In contrast, the 1957 novel *The Glass Bees* by the German literary author Ernst Jünger demonstrates a surprisingly posthuman take on the swarm as a technological community of the multitude, which is imagined beyond gender binaries and sociocentricity. In addition, my discussion of drone swarms in social movements sheds light on the swarm beyond its military connotations.

I further explore the nonhuman and zoological dimension by analyzing the use of drones for virus tracking during the COVID-19 pandemic. Chapter 7 shows how drones served as media to surveil lockdowns, enforce social distancing rules, monitor crowds, and even spray disinfectants. Artists have engaged with the pandemic drone, and I focus on artworks that represent it beyond the aspects of policing and militarization: an independent movie—made in 2016!—that portrays drones as saviors in the midst of a global pandemic; a YouTube movie about a city in lockdown and the "empathic" gaze of a drone; and the phenomenon of COVID drone shows as mass ornaments in the sky.

Let me conclude my introduction with a few words about my methods and the general structure of the book. As should have become clear from the chapter outline, each individual chapter focuses on a different register of drone sensing. Thus each chapter helps to reveal the richness, multisensoriality, and heterogeneity of the nonscopic drone sensorium. For this reason, the book does not follow a linear structure whereby each chapter would serve as an argumentative stepping-stone for the next. Rather, the chapters constitute exemplary case studies of specific drone sensing modes. Each chapter is intended to demonstrate the multimediality of the drone sensorium and shed light on its different imaginaries of communities. Occasional historical excursions provide a trajectory and an intervention to rethink the spectacularizing "drone-o-rama"[50] narratives that one-sidedly celebrate drone technology.

The artworks I discuss throughout this book demonstrate that the drone is a multisensorial device, and they reveal its community-shaping power. In my eagerness to prove this tidea, my discussion of these specially selected artworks perhaps risks harming their uniqueness, complexity, and singularity. I am aware of this. But this book is not intended as a hermeneutic or philological study of aesthetic works. Rather, my goal is to argue that drones are multisensorial assemblages that in turn make us rethink narratives and visions of communities and technology. At times the artworks may be forced to pay a price for this conceptual framing of my book. To analyze artworks is a scholarly method for me: through art analysis I gain new insights into the epistemological power of technology. In my defense, any aesthetic analysis involves a minimization of aesthetic uniqueness, since taking up a particular interpretative stance is inevitably also a decision to leave something unobserved. My readings of the artworks nevertheless strive to make the reader aware of this dilemma, and I make an effort to highlight the multitude and complexity of the drone artworks' interpretative layers.

The artworks discussed in *The Sensorium of the Drone and Communities* create a critical awareness regarding what we can learn from drones in respect to community models. Thus I do not seek to answer the question of whether drones are good or bad. Instead, I treat drone art and aesthetic drone sensoria as prisms through which we can observe our communities with remote sensing technology. This is an important—but unprecedented—undertaking in drone research, since drone technology and its field of vision are often conflated with a Western martial gaze. While this conflation is often accurate—this book does not aim to construct a techno-optimistic narrative of the "good" drone—it is nevertheless crucial to recognize the plenitude of the drone's different aesthetic sensoria and community visions.

I   The Sensorium of the Drone

1 The Sensorial Experience of the Drone and Communities

In the ancient world, a gorgon was not someone you would want to engage in a staring contest. According to Greek mythology, the gorgons were the sisters Stheno, Medusa, and Euryale, monsters with venomous snakes for hair; Medusa could literally petrify anyone who dared to gaze upon her. Since 2011, a gorgon has been working for the Central Intelligence Agency and the US military. It is named Gorgon Stare, and it was brought to life by military engineers in the labs of the Defense Advanced Research Projects Agency.[1] This modern gorgon shares its ancient relatives' fatal powers of vision: a spherical array of nine cameras is attached to an aerial drone armed with Hellfire missiles. Designed as a wide-area surveillance technology, Gorgon Stare is able to capture motion images of a whole city. When it is combined with its brother, the surveillance technology Argus-IS (another telling name), which contains over a hundred cellphone-like cameras, the two can patchwork together a megastream of images and feed them into networks of ground control stations. Gorgon Stare proposes a superhuman point of view from above, and its mythical name casts its visual power as legendary, archaic, and eternal.

Not only is drones' Olympian potential hailed by the US military's apparent taste for the classics, but researchers too have tended to fetishize the drone as a scopic regime of aerial super vision.[2] This book intends to undo the myth of the drone's hypervisual powers. The drone is not an all-seeing eye in the sky, but a sensorium that is capable of many different modes of nonscopic sensing. I understand the sensorium of the drone as a complex, multimedia, synesthetic, sensing assemblage, in which the human agent is enmeshed with the drone's technical apparatus. Drone sensoria can sense in many more ways than the notion of scopic regime suggests, as drone sensing can be blurred, flattened, rasterized, three-dimensional, volumetric, or swarmlike. To be sure, the visual is often the primary mode of drone sensing, but it does not necessarily take the form of verticalized and omniscient vision. In addition, these different modes of seeing often connect to other sensuous registers, such as sound, touch, smell, and thermal sensing. As I explained in my introduction, the big-picture idea behind this book is that these aesthetic drone sensoria construct imaginaries of communities. In research about drones, communities have mainly been discussed in military contexts and seen as products of the scopic regime. In contrast, I argue that aesthetic drone sensoria can construct communities on the basis of decentralized, networklike, and fluid sensing processes. In order to develop this argument in more depth, this chapter lays some historical and theoretical groundwork, defining the notions of *scopic regime*, *sensorium*, *assemblage*, and *community*. It begins by tracing the cultural narrative of aerial omnivoyance, which frequently overlooks the nonscopic aspects of aerial sensing.

The Scopic Regime

In the field of visual studies, the term *scopic regime*[3] expresses the idea that not only *what* we see but also *how* we see is historically rooted and conditioned by different times and places. The scopic regime is a mode of visual apprehension that is culturally constructed and prescriptive; it is also historically variable and can exist within a single cultural and social formation.[4] The notion of the scopic regime often plays a role in analyses of discourses of sovereign power, particularly in the context of warfare.[5] Etymologically, the Greek word *skopos* implies a direct connection between watching and waging war, as it can refer to both watcher and target.[6] The term *scopic* thus becomes synonymous with the martial gaze. Following Paul Virilio, we know that military power is more than just munitions, troops, and weapons.[7] War technology is also about vision: cameras employed during warfare, images mediated from the battlefield, perceptions of war by soldiers, victims, and the public. The aerial view often embodies the imperial worldview, serving modern forms of territorial and military power.[8] The concept of a universal, all-seeing perspective has been thoroughly incorporated into colonial, state, and military modes of organization, management, and planning.[9] Scholars of military drones frequently use the term *scopic regime* to characterize the way combat drones perceive their targets.[10] According to Derek Gregory, drone vision embodies a "militarized regime of hypervisibility"[11] that implements a vertical and synoptic view of the surveilled area. In research on military drones, the connection between the drone's scopic regime and its vertical execution of sovereignty has been widely discussed.[12]

This verticalized optics of the military drone is often projected onto the civilian drone, with researchers suggesting that the former's scopic regime spills over into the latter.[13] The *politics of verticality*, to use Eyal Weizman's famous term,[14] has intruded into the domestic realm in the contexts of leisure, agriculture, border security, and pipeline monitoring. As Roger Stahl notes, "The diffusion of drone vision into commercial space is also a symptom of a larger shift in cultural and political discourse that has recoded domestic space as a sphere of military concern."[15] Similarly to their older and bigger military siblings, civilian drones use a wide array of sensor technologies to make us see things that our natural eyes cannot. While there are noteworthy technological differences between military and civilian drones, there certainly are some "family resemblances,"[16] to adopt a term from Ludwig Wittgenstein. It is a rather common view in media studies that consumer culture's media technologies are a product of the closely intertwined development of technologies of war.[17] But that does not necessarily mean that civilian drones sense the world in the same way as military drones, or that they abide by military drones' scopic regime of violence.[18] In order to better grasp the civilian drone's nonscopic mode of sensing, let us now make a short excursion into the cultural narrative of the aerial view.[19]

The Aerial Gaze: From God's Eye to the Panopticon

The view from the sky has been associated with godly vision and supreme power since antiquity. The eyes of Apollo were thought to have sovereign powers, as his gaze was embedded in a cosmic vision of the globe.[20] This association of the eye in the sky with a godly all-seeing gaze is also reflected in Christian mythology. St. Augustine wrote that the eye of God was able to see into the human psyche,[21] and in Christian visual culture one finds many images, illustrations, and paintings that convey this idea of an omnivoyant eye from above.

Consider the print by the French illustrator Jacques Callot, who, in 1628, created a series of illustrations intended to introduce novices to monastic life. The illustration shows a rural landscape, with a sheep pen guarded by a dog, and some village settlements in the background. The most eye-catching feature is the vertical rod with an open eye at the top, which symmetrically divides the illustration. It embodies the eye of God and may remind us today of a closed-circuit television camera. At the foot of this biblical rod sits the sheepdog, watching its flock. The image thus implies a clear hierarchy in which the eye of God (at the top of the rod) observes the dukes (dog) who guard the people (sheep).[22] The composition represents a vertical power relation, since the linear perspective down from the rod suggests a geometrical constellation of power and social order.

René Descartes's writings on vision shifted this godly ocular power into the rational mind of the subject. By experimenting with an ox's eye, he used the model of the camera obscura to infer insights into the realm of human vision. After removing the eye's outer membranes, Descartes hung its lens in a darkened room that had only a single-point light source. When he pushed a curved disk (a piece of paper, or an eggshell) into the inside of the lens, images of the outside world were projected onto the disk, as if onto a retina. This experiment was considered to have demonstrated conclusively

Figure 1.1
Jacques Callot, *The Vigilant Eye* (1628). © Kupferstichkabinett, Staatliche Museen zu Berlin.

that humans did not perceive the outer world directly, but rather looked at inner images: a proof that the images perceived by the human subject were not sheer representations of the outside world but projections of the rationalized mind.[23] Descartes's model of vision thus suggested a "reduction of perceptual space to mathematical and homogeneous space, with its understanding of vision as monocular, static, fixed and immediate, distant and objectifying, purely theoretic and disincarnated."[24] According to Descartes, vision was a product of the mind; for him, that also meant that it was subject to mathematical and geometrical rules and the processes of rationalization.[25]

In the course of modernity, this rationalization of vision became an embodiment of secular control. The power of the godly all-seeing eye was transferred to the human subject, whose gaze could master, rule, and steer the world. This modern rationalization of vision can be found, for example, in urban planning, such as in the eighteenth-century architectural sketches of Claude-Nicolas Ledoux, who developed (and partly realized) a town for workers.[26] Ledoux placed the mayor's house at the center of the town, from where the mayor could oversee the community. This architectural construction also resonates with Jeremy Bentham's prison plan. As Michel Foucault has shown, Bentham's panopticon suggested a secularized eye of God executing omniscient surveillance.[27]

These few examples already demonstrate that the view from above is frequently embedded in a narrative of vertical control and power.[28] As Peter Adey puts it: "The genesis of modern systems of control is to see the aerial view within wider scopic regimes."[29] From the eye of God to the panopticon's watchtowers, the aerial gaze exerts violence and aggression.[30] But although this connection between the gaze from above and sovereignty certainly describes the drone's biopolitical power, I will show that it is not the whole story about the aerial and the drone. The drone can also sense the world according to different, nonscopic, modes. Everything changes when a pigeon takes the picture.

The Aerial Gaze beyond the Scopic: Pigeons as Remote Sensors

The German inventor, apothecary, and businessman Julius Gustav Neubronner strapped small automatic cameras to the breasts of homing pigeons. The pigeons then took aerial photographs around the city of Dresden and the town of Kronberg. The first filming pigeon was airborne in 1903, and Neubronner patented his invention in 1908. Similarly to drones, homing pigeons represent a dual technology of remote sensing. Neubronner's pigeons were used to deliver medication, and they also became a tourist attraction in Dresden. The public could observe the pigeons, and the photographs the birds "took" in flight were turned into postcards for sale. In addition to such entertainments, homing pigeons flew missions in warfare for aerial reconnaissance. Some pigeons even received medals for military service. But what happens to the aerial view when the camera is attached to a pigeon? Let us take a look at an image taken by Neubronner's pigeons in flight.

The pigeons' pictures do not abstract the aerial view by showing clusters, patterns, and surfaces (a phenomenon I discuss in chapter 4 in relation to flattened sensing, aerial balloons, and drones). The pigeon images are rather detailed, and they are shot from a low angle, showing aspects of street life and architecture. The vertical dimension defines these photographs to a certain degree. After all, they embody the bird's-eye view (although the pigeon does not see with its breast and has its eyes on each side of their head). Thus this vertical axis does not revolve around a single point and does not represent the panoptic perspective of super vision: pigeons do not see according to the scopic regime.[31] Instead, the vertical axis appears to be unhinged, fluid, and amorphous, since the pigeon camera tilts the perspective into a street-view level. Although the line of the horizon is present, it does not provide a stable frame for the picture. There is no vanishing point to order and centralize the perspective. The slanted horizon evokes disorientation, and in another photograph of Kronberg Castle we even see

Figure 1.2
Julius Gustav Neubronner's pigeon photography. Street views of Frankfurt (ca. 1908). Public domain.

the pigeon's wings. Aerial images taken by pigeons are not very reliable, as pigeons can get lost, fly off course, or decide to peck a worm rather than sail around with a camera.

More importantly, this pigeon perspective is *nonhuman*. As Joanna Zylinska notes in her discussion of nonhuman vision, pigeons (like dogs, insects, and satellites) can articulate what Donna Haraway has called a partial standpoint.[32] Pigeons see "through the eyes of the other,"[33] which in turn deconstructs the ocularcentric narrative and epistemology of vision. I follow Zylinska in seeing pigeon vision—and as I will show, drone vision—as an "ethico-political pointer"[34] that allows humans to reflect on the limitations of the human view, and which can be constructive for thinking about the imaginaries of communities the drone can suggest. The example of the pigeon as a remote sensing technology can help us to understand the nonscopic form of aerial vision that is also intrinsic to the sensing modes of the drone.

Drones often have bird names—the hawk, the lark, the hummingbird, the raven—and this points to the closeness of drone technology and the zoological (an aspect I further explore in chapter 6). When developing drones, engineers orient themselves according to the flight behaviors of birds. Drones and pigeons can both be used in dual contexts (military and civilian), and I can imagine that the award of a medal to a pigeon would have been just as contested as the US Air Force's award of honors to drone pilots today. Moreover, drones and pigeons alike can become wayward, are highly dependent on winds and weather, and can easily fly off course or go rogue.[35] But most importantly, drones can produce nonscopic aerial images that are rather like those retrieved from Neubronner's remote sensing pigeons. Take, for example, the combination of drones and GoPro cameras. GoPros are small, robust cameras that can be mounted directly onto the filming subject. Often used to record adventure and action scenes in leisure and professional filmmaking, they enable a first-person perspective. They can record a first-person view of a skiing tour or surfing trip, suggesting an immediate, subjective, and immersive experience. GoPro cameras can also be attached to drones, widening the latter's field of vision, because the drone camera is no longer facing downward but can—like the pigeon— suggest a tilted and widened view: an embodied bird's-eye view.

This image shows an aerial view shot by a drone flying over Paris. Similarly to the pigeon photographs, it displays a slanted view of the city. The nonscopic image suggests the embodied first-person view of the flying

Figure 1.3
Paris seen from a drone with a GoPro camera. Video still from Vigibot, *Drone Survol Paris La Défense: FPV GoPro Flight: Vidéo Unique En France Et Pour Longtemps!* (May 23, 2013). © Vigibot.

drone, but instead of seeing the bird's wings, as we do in Neubronner's photographs, in this image we see the drone's sensors at the bottom margin. GoPro drone videos often portray cityscapes from above, but they also record flights through abandoned buildings, beaches, and mountain gorges. Frequently these videos are accompanied by music, which in turn heightens the sensorial and immersive experience. I will discuss this embodied (nonscopic) drone sensing at more length in chapter 2, which analyzes hobby drone cultures, dronies, and drone racing. For now, it is important to note that if we look back at analogue aerial sensing technologies such as Neubronner's pigeons (and aerial balloons, as I will show in chapter 4), we become aware that aerial vision does not always have to be defined as a scopic regime of verticality. Our excursion with the homing pigeons should inspire us to understand the drone not only as an all-seeing cyclops eye in the sky but also as a sensorial assemblage with heterogeneous modes of human and nonhuman and sensing.[36]

The Sensorium of the Drone as a Human–Machine Assemblage

Vision is undoubtedly the key mode of drone sensing. All drones have cameras, and probably all drone pilots have eyes. However, drone vision is always entwined with other senses and systems of perception. There are drones that can detect smells: the so-called sniffing drone, which can sense gas leaks and react to sulfur and methane emissions.[37] Drones can specialize in sonic sensing, identifying sound signals and auditory patterns. And there are drones loaded with software that registers kinetic movement patterns with infrared and thermal sensors.[38] Drones also supersede their status as optical-only media because of their capacity for datafication and sensing via data.[39] All in all, the drone cannot be reduced to a technology of seeing alone; it has to be understood as a "synesthetic medium,"[40] a "more-than-optical"[41] interface that connects different modes of sensing, agents, environments, and machinic systems. The drone sensorium is a complex, multimedia, synesthetic, sensing assemblage in which the human agent is entangled with the drone's technical apparatus. Before I expand on the notion of the assemblage, let me discuss briefly how I understand the notion of sensing.

In broad strokes, sensing embodies neurophysiological processes based on stimuli, such as visual, acoustic, kinetic, or olfactory signals that affect the subject. In addition to these external stimuli, the sensing process also has a prediscursive layer that neuroscientists call *embodied simulation.*[42] In a nutshell, this involves the activation of an inner, prereflective body schema (which is different from the conscious body image) as a prerequisite for perception. Thus sensing is not only about registering external reality but also about resonating within one's own inner body schema on a neural level. This phenomenon of the body schema leads to interesting questions concerning machinic sensing, as I show in chapter 2 (on drones and embodied sensing) and chapter 7 (on drones and empathic sensing during the COVID-19 pandemic). Moreover, sensing is always defined by historical contexts and technological developments. As Walter Benjamin states, "During long periods of history, the mode of human sense perception changes with humanity's entire mode of existence. The manner in which human sense perception is organized, the medium in which it is accomplished, is determined not only by nature but by historical circumstances as well."[43] Human sensing modes are subject to historical, social, and technological processes, such as urbanization, the invention of new transportation systems, and the rise of mass media, which all change the ways in which the subject perceives reality. Jonathan Crary describes this historical construction of seeing in his analysis of nineteenth-century observation techniques. His work emphasizes that the observer is always embedded in a set of (historical) conventions and conditions, and thus the observer "sees" according to preexisting possibilities.[44]

Human sensing is affected, shaped, and formed by technology—and vice versa. In the sensorium of the drone, this means that the human agent and the drone apparatus are entwined. The sensorium of the drone is not entirely machinic but represents an "entire perceptual apparatus."[45] The drone is thus more than a prothesis that helps us to see, and it is more than an extension that widens and sharpens the human process of sensing. Like the human subject, the drone sensorium can sense on its own, to a certain degree. Drones can steer their own sensing processes, select the sensed data, interpret them according to algorithmic programs, and select and operate with them. However, their ways of sensing have to be perceived as being more metaphorical, as drones do not have the capacity to understand and experience the world hermeneutically. A drone and its sensorium cannot intuitively grasp the world by raising questions, projecting meanings, or reflecting on its own perceptions. Most importantly of all, the drone always senses together with the human. Although the synesthetic drone sensorium is a remote sensing technology, it always needs a human agent somewhere in the loop of perception.

This is the case in military contexts, where the drone—as an "unmanned aerial vehicle" (UAV)—is not the main part of the operation but comprises an immersive synthesis of sensor technologies and human agents.[46] Drone operations require human collaboration to achieve "situational awareness," that is, knowledge about a situation within an area of operation based on sensor data.[47] The human agent is part of the assembly of hardware, sensors, data, algorithms, and agencies. In the context of civilian drones too, it makes sense to talk about a sensorium that combines human and non-human sensing elements. Like their military relatives, civilian drones are technical systems that rely on interfaces such as remote controls, smartphones, the Internet, satellites, and airspace regulations, but also on the pilot's hands and on the pilot's actions and decisions.[48]

The sensorium of the drone thus suggests an interactive human–machine assemblage. In *A Thousand Plateaus*, Gilles Deleuze and Félix Guattari use the term *assemblage* to describe an association of heterogeneous elements.[49] For them, the assemblage is an entry point into a theory about the body and its relationship to society. It defines the body not as a stable and fixed entity but as fluid, exchangeable, multifunctional, and constantly defined by new processes of connection and disconnection. This dynamic conception of the assemblage as connecting disparate elements is productive for characterizing the drone sensorium. The drone sensorium as an assemblage suggests neither that the subject is external to the remote sensing technology nor that it is its controlling master. Rather, the human agent is entwined with this technology in a relationship of mutual dependence and independence.[50] We might also grasp the drone assemblage through the ideas of Gilbert Simondon, who develops a general phenomenology of technical objects. According to Simondon, technical objects become "mediators between man and nature,"[51] positing a correlative existence between humans and machines. The interconnectedness of the human agent with the machine illuminates the idea that technology is not just about hardware but also about the human that engages with it. By virtue of this human entanglement, technology becomes a mode of constructing and making the world.

Technology assemblages certainly did not start with the invention of the drone. They have existed since the early days of humankind: the Neanderthal embodies an assemblage with a flint. Martin Heidegger's discussion of the sacrificial bowl as *techné* is a debate about a human–technology assemblage.[52] Visual technologies, such as eyeglasses, telescopes, and microscopes, form assemblages, as they have the prosthetic purpose of enhancing human vision.[53] I could list many more examples. The drone is by no means a game changer. However, the drone assemblage does intensify the interdependence of the human and the technical object, because the drone as sensorial technology can attain a higher degree of decision-making autonomy.

The question of the drone's autonomy is important in current debates in philosophy, international politics, ethics, and science and technology studies.[54] Artists and visual culture scholars also address questions of drone autonomy, and in this context the discussion of operative images has been decisive.[55] The term *operative image* stems from visual artist and writer Harun Farocki's work on remote-controlled missiles in his video installation *Eye/Machine I–III* (2001–2003).[56] Images made by drones can also be categorized as operative images, since they are often first "seen" only by machines, are sorted by algorithms and artificial intelligence (AI) programs, and gain operative power with regard to targeting.[57] Collected in big-data archives, they are no longer defined by representation; rather, they trigger actions and decisions as they select, order, and demarcate targets.[58] In this respect, operative images gain a type of operative autonomy, as they process and interpret data on their own.[59] However, at some point human agents are present in this process: they design the machine reading processes, and they interpret the data. Humans may not be the masters of all these processes, but they are, as Gregory has put it in relation to military drones, part of the "kill chain."[60] The term *kill chain* indicates the complexity and intricacy of the military operation. A drone strike consists of several phases, such as surveillance, identification of the target, initiation of the attack, and destruction. In a drone kill chain, humans and machines (UAVs, computers, algorithms) work together and build a military operational network, extracting knowledge from the image streams the drone delivers. Hence human agents' interpretation of the data is not neutral, since neither the humans nor the image streams are neutral. There are no "raw data": the data are preselected and preordered by algorithms.[61] Although these operative images do operate, we cannot conclude that drones are autonomous machines that remove all human decision making. Rather, drones and their sensorial technologies have to be thought of as sensorial assemblages that entangle both human and nonhuman modes of sensing. Put differently, the drone embodies a form of cyborg.

Drones as Cyborgs

Long before high-tech drones appeared in the sky, Haraway discussed human–machine assemblages in her feminist critique of science, aided in particular by the figure of the cyborg. For Haraway, the cyborg is a "cybernetic organism, a hybrid of machine and organism, a creature of social reality as well as a creature of fiction."[62] AI, stem cell research, nanotechnology, and the Internet are areas that breed cyborgs. Even the cellphone can be seen as a device that creates cyborgs: it can attain the status of a physical organ, and when it is not available some users display the "missing limb" symptoms observed among amputees.[63] (I know this from personal experience since I am currently living with two teens and their cellphones.) The cyborg is a human–machine assemblage, with an element of the nonhuman vision that we have already discussed in the context of homing pigeons. The cyborg does not see all; its view is partialized. Haraway attacks the idea that the eye as a human organ, together with its technical enhancements, embodies a metaphor of discursive clarity and human mastery.[64] In opposition to this ocular-centric worldview, she argues for another type of vision, in which the subject (or what is left of it) is decentered. She suggests a model of nonhuman partialized vision, such as that of a cyborg, a dog, a satellite, or a pigeon. In these instances, there is no Archimedean point from which one can see everything; the partialized view does not allow us to play what Haraway calls the "God trick."[65] In this partial view Haraway sees an opportunity to formulate an (alternative) concept of objectivity. Since the partialized view, or "situated knowledge,"[66] can no longer be subsumed by either universalism or relativism, it has the potential to articulate a model of objectivity beyond dualistic thinking.[67] Haraway embraces the cyborg in this context because it can question the dichotomies between human and machine, male and female, organic and nonorganic.

For me, the drone too is a type of cyborg.[68] But I am less interested in a philosophical argument about objectivity than I am in Haraway's critique of dualism. I propose that we need to see the drone as something that transcends binary oppositions, such as that between machine and human, or between technological object and subject. Rather than labeling the drone "dehumanizing" and thereby reinforcing the opposition between human and machine, I believe it is more productive to understand the drone (and its sensorial power) in terms of its antidualistic features. If we do this, the drone's power is no longer solely grasped by the scopic regime of verticality, which suggests that there is one unified authority behind the eye of the drone camera. Instead, we can say that the drone assemblage also works as a horizontal network connecting machines, data, and human agents in a platform of dynamic and amorphous interfaces.

Alexander R. Galloway and Eugene Thacker have discussed the power of networks in relation to sovereignty, noting that networks "exercise novel forms of control that operate at a level that is anonymous and nonhuman, which is to say material."[69] This network power also characterizes the drone. Such power does not follow a vertical axis. It is decentralized, and its decisions rely on multiple actors, defying strict boundaries, territorial/temporal limits, and coherent individual selves. Researchers on military drones have investigated their network power and shown that the drone's violence lies in precisely this dissociated, decentralized, and flattened network configuration.[70] Civilian drones execute network power too, and chapter 3 discusses this in greater depth. However, I do not intend to reduce the drone to the notion of the network, as this notion describes only one aspect of its sensorium. For me, it is important to understand the drone as a cyborg primarily because doing so reveals that drones can overcome dualistic and binary modes of thinking (for better or worse). A drone cyborg as a technosensual assemblage presents an epistemological prism through which to gain new insights about human–machine relationships and communities. On the one hand, my book allows the drone (in part) to be more playful than harmful, drawing on Zylinska's idea of the creative, life-making potential of nonhuman photography. Zylinska sees nonhuman photography as "offering the notion of 'nonhuman vision' as an alternative vantage point from which to understand ourselves and what we humans have called 'the world,' in all its nonhuman entanglements."[71] Nor do I comprehend the drone as merely a killer robot; rather, it is a sensorial and technological assemblage that enables new forms of relating between humans and machines, and most importantly opens up new understandings of communities. On the other hand, however, this creative aspect of the drone sensorium is not a free pass to celebrate and fetishize the technology. It is paramount to discuss both the affirmative and the critical potentials of the drone's nonhuman perspective.

Aesthetics and Communities

As already outlined in my introduction to this book, I think that the realm of aesthetics can set free the creative sensorial experiences of the drone, shed light on its epistemological power, and experiment with new imaginaries of communities. Jacques Rancière's aesthetic theory is productive here. For Rancière, aesthetics can disrupt, provoke, and reflect, and it can do so because it manifests itself through noninstrumentality and nonfunctionality: "The aesthetic regime of the arts is the regime that strictly identifies art in the singular and frees it from any specific rule, from any hierarchy of the arts, subject matter, and genres."[72] Aesthetics can redistribute "the sensible"— all that is commonly sensible, sayable, and visible in society[73]—by disturbing its acceptance and creating dissensus.[74] Rancière frequently makes use of the notion of aesthetic sensorium in this context: "Aesthetic experience is that of an unprecedented sensorium in which the hierarchies are abolished that structured sensory experience."[75] The experience of art can enable a visceral, affective, prediscursive form of sensing—a type of aesthetic unconscious— that is in negotiation, dissensus, and disagreement with the sensible.[76]

For Rancière, aesthetics is a privileged realm in which to experience the organization of the sensible world. Thanks to art's autonomy, freedom of purpose (*Zweckfreiheit*), and noninstrumentality, it represents a medium through which—in theory—one can experiment with the uttering of different opinions, judgments, and tastes, without being constrained by religious, political, or economic codes. This aesthetic experience of dissensus offers a trajectory into the political dimension of art. This is not "political" in the sense of party politics, state politics, or law enforcement. Rather, it has roots in German Romanticism and Friedrich Schiller's idealist notion of the aesthetic state: the political conceived on the level of language, images, music, and poetry. Through the experience of art, one can practice dissent, dismantle hierarchies, and exercise freedom. Rancière thus connects aesthetics with a discourse of political critique that receives its power precisely from its inherent noninstrumentality. This is of course a contradiction, since at the very moment when aesthetics and the political interconnect, art loses its autonomous status. But we should not understand Rancière as suggesting that art represents social utopias or illustrates political ideologies: the political lies in the aesthetic texture of the artwork, and in its aesthetic power of "common" dissensus, resistance against the sensible, and critique. The connection Rancière draws between the artistic and the political via a common aesthetic experience reveals how the sensorium of the drone can be a generative matrix for new imaginaries of communities.

Sociological and philosophical discourses often define community in terms of sharing or having something in common. For example, a religious community shares a common belief and common place of worship, a regional one a specific language and geographical space, and a national one a common foundation myth. For Ferdinand Tönnies, *community* is a counterterm to *society*: the smaller, more personal *Gemeinschaft* in opposition to *Gesellschaft*.[77] Tönnies sees community as a more affective, personal, or intimate social configuration than civil society. This view of community as more authentic than the state has determined many conceptions of community in the history of political thought and has a long history of political instrumentalization. Think of the *Volksgemeinschaft* during National Socialism, which relied on an essentialist and racist definition of its community members; or of communitarianism, in which communist ideology defined the authentic community on the basis of productivity and political ideology.

Jean-Luc Nancy reworks these concepts and puts the notion of community into a broader politico-ethical and cultural context.[78] His aesthetic theory criticizes *community* as a dominant Western political formation that is founded on exclusionary myths of national, racial, and religious unity. He seeks to undo these essentialist notions of community, and to accommodate more fluid and dynamic forms of being-in-common and dwelling together in the world. Communities are not about fixed conditions; rather, Nancy understands communities as relational social manifestations, in negotiation with and openness to each other: "Community is what takes place always through others and for others. It is not the space of the egos—subjects and substances that are at bottom immortal—but of the I's, who are always *others*."[79] The members of the community are not individuals that can be subsumed into a totality. They are singularities ("I's"), a group of "others" that are finite and mortal. Nancy's notion of community goes against the idea of immanence as something eternal, permanent, higher, or even divine ingrained in the emergence of a community.

Community for Nancy is an attempt to think a form of being-in-common that has nothing to do with fusion into an identity. For him, community emerges from the shared lack that evolves when modern ideologies turn the notion of community into something violent. This bereavement—which Nancy also calls the moment of inoperativity—is constructive.[80] In order to facilitate this constructive power of loss, the realm of aesthetics is key, as it enables the moment of (nonidentical) sharing in the inoperative community.[81] Specifically, poetic language, mythic speech, and its power of dissemination can generate these formations of inoperative communities. Nancy's focus on the community-shaping power of aesthetics complements my intention to show how drone art—and its creative negotiations of the drone sensorium—can construct imaginaries of communities. Like Nancy, I see the aesthetic realm as key to the experience and envisioning of imaginaries of communities. Contemporary drone art almost always has a vision, critique, and reflection of communities at its core.[82] Nancy is an important point of departure because his notion of community provides a theoretical scaffolding for me to experiment with new imaginaries of communities in respect to the sensorium of the drone.

In doing so, I reach beyond the phenomenological and deconstructivist context in which Nancy embeds his notion of community, and I connect his idea of the inoperative community to philosophical tenets of posthumanism—a connection that has not previously been made in research on technology and communities, and from which I hope to model a form of nonidentical community in light of remote sensing technology. Rosi Braidotti's thinking from difference in regard to communities, as well as her focus on and embrace of technology, is vital here. In her work on the posthuman condition, Braidotti observes that the question of what and who is human has become a central concern of our time in light of technological developments, such as robots, AI, drones, and algorithms.[83] The omnipresence and ongoing development of smart technology fundamentally interrogates and redefines perspectives on the human. In the contexts of what Braidotti sees as the coming Sixth Extinction and the ongoing Fourth Industrial Revolution, the human subject and its status as the "crown of creation" are in deep crisis. She is certainly not the first to make these observations, but her constructive interpretations of technology have been productive for my investigation of the drone sensorium. Braidotti's theory does not fall into a Luddite technological nostalgia, but instead finds life-affirming, vital, and constructive aspects to technology. I try to do the same: despite the lethal power of the drone sensorium, I investigate what Braidotti calls the "positive potential of the posthuman convergence."[84]

Braidotti rethinks the anthropocentric concept of the subject in the face of current challenges. As an embodiment of the masculine dominance of thought, the authority of the universalist, unified Enlightenment subject is fading. At stake instead is a model of relational subjectivity that is no longer rooted in the unified subject—a form of subjectivity that is multifaceted, differential, and even paradoxical. In contrast to some similar postmodern models, Braidotti argues that "nomadic" subjectivity does not celebrate an "anything goes" ethical relativism: "Non-unitary visions of the subject do not result in cognitive and moral relativism."[85] Being posthuman does not imply being inhuman, but rather entails an opportunity to think the human into the picture in a complex posthuman world. Moreover, this new form of subjectivity does not pertain exclusively to the human. Braidotti understands "posthuman subjectivity as an ensemble composed by zoe-logical, geological and technological organisms—it is a zoe/geo/techno assemblage."[86]

This dynamic, fluid, and nonhuman-centered vision of subjectivity embraces nonhuman elements, such as machines, and in doing so it generates possibilities to rethink communities. Indeed, the decentering of the subject does not push the human and its social and community-building features out of the picture. Braidotti finds new "possibilities of bonding, community building, and empowerment,"[87] and she suggests that "posthuman subjectivity expresses an embodied and embedded and hence partial form of accountability, based on a strong sense of collectivity, relationality and hence community building."[88] Like Nancy, Braidotti bases communities on difference and singularities, and she searches for forms of social bonding and relations that are fluid, dynamic, and nonexclusionary: a social bonding described by the term "We-Are-(All)-In-This-Together-But-We-Are-Not-All-One-And-The-Same."[89]

In sum, against the theoretical backdrops of Rancière's aesthetics, Nancy's notion of community, and Braidotti's ideas about posthumanism, this book goes in search of new forms of communities associated with drone technology. But I have also chosen the notion of *community* (rather than, say, *society* or *social network*) because it highlights that there is something we share with drones. This sharing is not necessarily an intersubjective activity in which something is given out and communicated exclusively from human to human. Instead, it is a sharing with the machines. The currencies of this sharing are the images, signs, signals, and bytes that the drone assemblage collects. These data are organized, selected, and ordered by the algorithms, software programs, AI, and human agents that distribute, exchange, send, receive, post, and circulate them. Sharing embodies processes of networking, data flows, AI, and information exchange between agents and machines, as well as among machines.

Thus drones and their agents share data under very different premises and contexts. As the following chapters show, the sharing of drone data can indicate giving to, or even caring for, a community. The drone not only provides intimate connectivity to the community but is also bodily and intimately attached to the human. For example, we find this personal community around drones in their use by protesters and environmental activists. In these communities, the sensorium of the drone and its data sharing can create new forms of human bonding beyond social divisions, such as planetary communities that are no longer human centered or exclusively anthropomorphic. The drone can suggest communities that are nontotalizing, relational, dynamic, fluid social formations beyond closed unities.

But the dialectics of the drone and its sharing activity can also take another turn: drone sensing can be devoid of mutual exchange and take data away from people. Drones (like smartphones, cameras, and sensor interfaces) have radically increased data mining, leading to a shift from the surveillance society to a society controlled by "always-on network sensors."[90] In this form of surveillance, there is no interactive sharing of data. Rather, the surveilled become anonymous units in a big-data pattern. The drone is a key figure in this ubiquitous, always-on, sensor-based monitoring and control.[91] Thus the drone as a datafied sensor affects communities in destructive and dehumanizing ways: precisely by virtue of the connection between machines and humans, drones can execute violence, power, and oppression and can function as instruments of data capitalism.[92]

II   The Body

2   Embodied Sensing and Cyborg Communities

For a long time, I thought that drones had nothing to do with bodies—except that they could kill. Aside from this cruel relation to the body, for me a drone was an apparatus that flew in the sky, detached from and remotely controlled by a disembodied pilot: an unmanned aerial vehicle (UAV). But then, in 2017, I went to the Intrepid Sea, Air, and Space Museum in New York City, where I visited an exhibit on drones. And there I encountered Volantis, Lady Gaga's monumental drone dress.

Volantis is a remote-controlled flying dress. It is powered by six lifting rotor units mounted in hexagonal formation, which give it the ability to hover above the ground. The white drone dress includes a small platform for Lady Gaga to stand on, shielded by a curvy female silhouette made out of light carbon-fiber plastics, which serve as a body piece and safety harness. Video materials at the exhibit documented Lady Gaga's spectacular performance as she wore (or better, flew) the dress during the 2013 artRAVE opening in Brooklyn. (Lady Gaga seems to have a knack for drones: think about her Super Bowl LI Halftime show with drones in 2017.)

As I stood in front of her excessive dress, the following thoughts raced through my mind: this Gaga drone is all about the body, and the pop star and the drone merge into a technosensual body assemblage. A baroque-style dressed-up cyborg gone feminist. Gaga is queering drone tech. Her drone dress engages with the body as a way of displaying and sensing, and this floating and hovering drone no longer peers down from the sky with a verticalized scopic gaze. I am not sure what Lady Gaga saw from her futuristic helmet, but I think it is safe to say that she did not have an experience of aerial omnivision. Likewise, the pilot who stood behind the drone dress did not see downward but, rather, merged with the drone in a first-person view.

The Gaga drone leads right into this chapter's key questions. In what ways does the body play a role in the drone sensorium? How can we understand the drone as a technology of nonscopic embodied sensing and telepresence? How do hobby dronists and artists negotiate this embodied sensorium of the drone? What communities do these technobodies of the drone envision? Can they shed light on alternative symbiotic human–machine communities? Or do they naturalize their capitalist legacies, namely, the surveillance power of tech companies and the military industry? In order to find answers to these questions, I first explain the idea of embodied sensing and its telepresence effects in a media-historical context. By focusing on the nineteenth-century panorama and telephonoscope, I establish the drone as a member of the telepresence media family. I then proceed to discuss dronies, drone racing, and the work of contemporary drone artists. I show that the drone as an embodied sensorium can evoke effects of telepresence, and in doing so can create new imaginaries of machine–human communities.

Embodied Sensing and the Effects of Telepresence: A Short Media History

Embodied sensing suggests a way of experiencing and perceiving the world as a process that embraces the whole body as an organism. It refers to the idea that besides the visual dimension, there are somatic, haptic, and sensor-motoric experiences that shape the human process of perception. In addition, the mode of embodied sensing rests on the idea of a body schema. This refers to how the body's position emits prediscursive sensory impulses that then inform further cognitive processes. Technology plays an important role in relation to embodied sensing: technological media can channel, facilitate, enhance, and substitute embodied sensing as they directly interact with our bodily positions, surfaces, and movements. Embodied sensing connects with Don Ihde's postphenomenological observations of how technology mediates, constitutes, and shapes human–world relations.[1] Technology is a medium that can embrace the materiality, situatedness, and relationality of experiences. In this respect, drone technology acts as a sensorium that is not detached from the outer or inner body or world, but rather is entangled and enmeshed with them. It matters not only how the drone as an object is held, touched, and steered, but also how it can merge with the viewpoint and affectivities of the pilot. The first-person view subsumes the pilot and the drone to create a new immersive space similar to the virtual spaces of three-dimensional computer games.

This capacity of the drone to act as a medium of embodied sensing interlinks with the idea of drones as devices that can trigger the effect of telepresence.[2] *Telepresence* refers to the sensorial condition or situation whereby a medium can make a human agent present elsewhere without that agent having to travel physically. As I write this book, in the midst of the COVID-19 crisis, telepresence has become a way to live and stay connected for many people. Some of us conduct our professional work via Zoom, FaceTime, and Skype. But the desire to be present at a place other than one's physical location has been around since well before the spread of coronavirus in 2020.

Media scholar Oliver Grau points to the mythical and occult roots of the human desire to leave the body, to overcome spatial and temporal boundaries like a god: "Robotics, telecommunications, and virtual reality feed into the history of the idea of telepresence—three areas that from their inception have featured repeatedly interpretations of the given technical stage of development charged with mythological/magical or religious overtones."[3] The sci-fi franchise *Star Trek* may have mastered the art of telepresence with its holodeck technology (a fictional device to simulate holograms), but telepresence has a far longer cultural history. Indeed, we can find early manifestations of it in biblical and other religious stories; for example, in the Christian story of Easter, when Jesus's soul left his body and traveled elsewhere. But those of us who are nondivine need a medium or technology that can transport us into another space without our actually going there. Images as cultural technologies (*Kulturtechniken*) can be media to induce presence in remote locations: "We can see this in the German word for image, *Bild* and its etymological Germanic root *bil*. *Bild* represents less the specifically graphic and more something that is permeated by an irrational, magical, and spectral power; which cannot be fully understood or controlled by the observer; an artifact that has the power to leave the body and possesses a life on its own."[4] Pictures often obtain the power of a fetish to interact with spirits, as in voodoo or other religious practices.[5]

For the purpose of my attempt to unearth family resemblances between the drone and telepresence technologies, nineteenth-century optical media are particularly interesting.[6] During Romanticism, artists, intellectuals, and scientists were fascinated by sensoria, robots, automata, and the idea of artificial life. Consider, for example, E. T. A. Hoffmann's "The Automata" ("Die Automate," 1814), or the uncanny doll Olimpia in his "The Sandman" ("Der Sandmann," 1817); or the spirit-homunculus in Johann Wolfgang von Goethe's *Faust II* (1831), a tiny, highly intelligent artificial human that lives in a glass bottle; or the Creature in Mary Shelley's *Frankenstein* (1818). These literary imaginings of intelligent machines and artificial humans arose amid the period's general interest in automata and androids. There were automata that represented artists, conjurers, and musicians; many were mechanical, and they often possessed music boxes.[7] These had been preceded in the eighteenth century by the famous mechanical duck designed by Jacques de Vaucanson, and by Wolfgang von Kempelen's chess-playing robot. As Grau emphasizes, these automata reflected the desire to expand the human body and play God, which also constitutes a key feature of telepresence technologies: "Through networking any number of robots or technobodies, telepresence facilitates the multiplication of possible spaces of experience."[8] Although these robots and artificial humans were not telepresence machines comparable to today's Zoom, they were informed by the telematic fantasy of extending human perception and intelligence into remote-controlled machines.

The nineteenth century is a cultural treasure trove not only of automata but also of media-optical devices with telepresence effects. In what follows, I focus on two such devices: the panorama and the telephonoscope. As with the hot-air balloon, which I discuss in chapter 4, I will consider these as distant relatives of the drone. Let us turn first to the panorama.

The word *panorama* is a neologism, formed by combining the Greek words *pan* (all) and *horama* (view).[9] A panoramic perspective thus entails a visual experience of a wide field, aiming to represent an all-encompassing view. But—and this is crucial to the panorama as a medium of embodied sensing and telepresence—it is not the overview that defines the panorama. Rather, it is the experience of immersion, the submerging of the viewer into a whole sensorial, simulated, quasi-three-dimensional space. As early as the 1780s, Robert Barker experimented with this perspective, creating large-scale paintings that suggested a 360-degree view and could immerse the viewer. Barker's first attempt to seek credit for his invention was not very successful, and the Royal Academy of Arts in London, founded in 1868 as a private institution that funded artists, did not acknowledge his paintings. Nevertheless, Barker patented his invention in 1787, and only a few years later he exhibited a

panoramic painting entitled *Cities of London and Westminster* (1792). He
showed this work in a specially constructed rotunda in his backyard on
Castle Street in London, for an admission price of one shilling. The exhibition of this semicircular painting (it was later extended into a full circle)
not only finally won him the acceptance of Sir Joshua Reynolds, a founder
member of the Royal Academy, but also launched the fashion for panoramas as public spectacles. Panoramas became popular attractions for mass
entertainment, accessible to the wider public for a small entrance fee. Visitors soon flocked to these optical sensations, not only in the British Isles
but also in Germany and other parts of Europe where the genre boomed.[10]

In order to see Barker's panoramic painting, the audience had to enter a
large rotunda and climb a spiral staircase to a balcony that was covered with
a huge, umbrella-shaped canvas, blocking the light from above. From this
platform, which was several meters high and could hold many people at
once, the audience had an all-round view, similar to the perspective from a
tower, ship's mast, or hot-air balloon. Since the panorama was built around
an elevated tower, it has often been connected to the panopticon's vertical,
colonial, and imperial gaze from above. While this panoptic was certainly
one aspect of the panorama, for my purposes it is important to note that it
equally offered a nonscopic sensory experience: it immersed the viewer in
a virtual reality reminiscent of today's digital computer games and three-
dimensional movies.[11] Since the viewer saw nothing but the picture, there
was nothing to break the illusion, and thus the panorama sought to achieve a
total illusion of reality. Jonathan Crary says of the panoramic perspective,
"One was compelled at least to turn one's head (and eyes) to see the entire
work."[12] The viewer saw a part of the representation, and by turning around
their own axis was able to get an all-encompassing view of the picture. This
roving gaze of the viewer anticipated filmic qualities.[13] No matter which
way the viewer turned, they would (ideally) never encounter a frame that
separated the panoramic picture from its environment or separated one picture from another within a sequence of movement. The panoramic experience thus suggested a framelessness: the viewer was completely embodied in
the panorama sensorium. In this way, the panorama simulated an effect of
telepresence, transporting the viewer to a completely different place without requiring them to travel physically. This effect was enhanced in Barker's
panoramic constructions by trompe l'oeil objects, such as artificial trees,
objects, and individuals that simulated reality. Stephan Oettermann states,

"The basic aim of the panorama was to reproduce the real world so skillfully that spectators could believe what they were seeing was genuine."[14]

My second example of a nineteenth-century telepresence technology to which the drone is remotely related is the telephonoscope. Unlike the panorama, the telephonoscope was never realized; it remained a technological fantasy. George du Maurier conjured it up in *Punch* magazine in 1878 as part of a cartoon series about fictional inventions that made fun of the famous inventor Thomas Edison. Du Maurier's telephonoscope resembled a very bulky videophone-cum-television. The cartoon shows a British couple talking to their daughter through an electric camera obscura mounted above the fireplace. We can see the pipe-shaped "receivers" that connect them audiovisually; the darkened room suggests an immersive, cinematic perspective. As in the panorama, the couple themselves are part of the illusional space. The text beneath the image reads as follows:

Paterfamilias (in Wilton Place): "Beatrice, come closer, I want to whisper."
Beatrice (from Ceylon): "Yes, Papa dear."
Paterfamilias: "Who is that charming young lady playing on Charlie's side?"
Beatrice: "She's just come over from England, Papa. I'll introduce you to her as soon as the game's over?"

Like today's Zoom calls, the telephonoscope takes its participants to other spaces and time zones. In doing so, it displays a fantasy of telepresence: its users are present in another reality, leaving their bodies behind and becoming immersed in the scene above their fireplace.

The drone belongs in this media archaeology of telepresence technologies because drones can project humans into other places and simulate their presence in spaces other than their true location. In a military context, for example, a drone pilot can sit in a container in Nevada and execute kill actions in Afghanistan. The civilian drones that I discuss in this chapter can also act as "presence machines," to use Maximilian Jablonowski's term, in the sense that they suggest the sensuous, affective presence of a remote agent via embodied sensing.[15] I will show that this is particularly evident with dronies and drone races, where drones stimulate an externalization of the body into a different time and space. But in experimental artworks too, the drone can evoke telepresence effects.

Figure 2.1
George du Maurier, "Edison's Telephonoscope," cartoon almanac for 1879, *Punch* magazine (December 9, 1878). © The British Library Board, Scala Florence.

Dronies and Drone Races: Extending the Body and Telepresence

Amateur drones are popular in hobby drone circles, also known as DIY (do-it-yourself) drone communities. These drones have become relatively affordable over the last decade and are easy to buy online. Around the globe, there are numerous hobby drone communities where pilots meet, fly their recreational drones together, post drone films online, blog about drones, and share technical know-how. For example, there is the Airvuz forum, a community of drone pilots and videographers who want to watch or upload aerial videos.[16]

Within this culture of hobby dronists, a new visual genre has emerged: the dronie.[17] The dronie is a short video shot by a drone. It first shows the pilot who is steering the drone, and then it zooms further and further away into the sky, until the drone's controller and point of origin almost vanish. Typically, dronies are taken in scenic environments, portraying ocean views, mountain vistas, or cityscapes, for example. They are also popular as records of weddings or other family gatherings, providing an alternative to the more traditional (static) family portrait or handheld video. Let us take a look at an example of a dronie.

This dronie is typical of the genre, as it is shot in a spectacular land-scape—in this case, the cliffs of Dorset in the UK. The pilot sits on a cliff and steers the drone away from the remote-control unit into the sky. The seem-ingly shrinking pilot thus describes the focal point of the drone camera. The dronie's backward zoom-out presents a principle of visual composition that challenges the vertical view: instead of offering a bird's-eye perspective onto the ground, the slightly tilted angle suggests a panoramic landscape shot. The dronie gradually opens and widens the surroundings and the hori-zon line.

From a media-historical point of view, the dronie bears some resem-blances to the nineteenth-century immersive mass medium of the pan-orama. Like the panorama, the dronie does not convey one verticalized panoptic master gaze. Indeed, the dronie seems to mock that perspective.

Figure 2.2
Dronie still from the cliffs of Dorset, UK (April 30, 2017). © jcourtial.

Where military drones can focus on a target on the earth, the dronie focuses away from the ground as it leaves its controller behind, and the pilot's gradual disappearance into the landscape inverts the target logic of the military drone view. The dronie's controller is the vanishing point, a figure that becomes minimized and eventually unrecognizable. The more the pilot disappears, the more their body seems to merge with the drone machine. Although the pilot is no longer visible, the dronie suggests their presence in the sky, as if the drone has taken the pilot's body and levitated it. Peter Adey has written about the levitation of the drone, and has suggested a change of perspective: instead of looking down from the sky, he focuses on the moment the drone takes off.[18] For Adey, the cultural and theological myth of levitation is key to the drone's fascinating power: its floating up into the sky recalls religious narratives of ascension.[19] Indeed, this idea of levitation links directly to the mythological and religious roots of telepresence, since the technology suggests a departure from one body and into another space or body. The dronie plays with this cultural and religious aspect of levitation. It literally elevates the pilot's body into the air, enabling it to be in the film, on the ground and in the air simultaneously.

This state of levitation is not informed by the scopic, verticalized master gaze of a dominating subject. The dronie fuses the pilot's body with the drone by making them into one intertwined technobody. This sensorial assemblage enmeshes the remote sensing technology with the pilot's body, a GPS tracking system, wireless radio frequency transmitters, and, ultimately, the technology of social media platforms and their multiple viewers. This levitated assemblage is no longer bound to the ground or confined by the natural eye; it is able to see and sense its environment bodily. Thus the dronie does not reiterate its military relatives' violent gaze from the sky; it hovers away from its controllers, even as it paradoxically takes them up with it.

Of course, the dronie tends to reiterate tourist aesthetics by filming spectacular landscapes and attractions. Like the panorama, which was a mass medium that exoticized the cultural other, the dronie can also be seen as serving the tourist industry.[20] Nevertheless, dronies have a different aesthetic than panorama travel shows. While panoramas often propagated nationalist and colonialist agendas of the state government, the dronie is a technology controlled by individuals and hobby videographers. Even when a dronie is shot in a tourist area, its strange zoom-away aesthetic does not correspond to a tourist aesthetic of sublime and grandiose representations of landscapes. In addition, the dronie is often highly subject to environ-mental factors, such as wind, weather, birds, and atmospheric changes, and this both diminishes human control and conditions what the drone will actually film.[21] Moreover, some dronies are not about traveling at all. For example, there is a dronie that shows a couple standing on a balcony in a skyscraper in Singapore; the couple wave as they slowly disappear into the aerial image of the city. Instead of romanticizing "exotic" places, the dronie highlights the capacity to merge and entangle the body with drone technology. The dronie pilots become part of the technological apparatus of the drone, as not only their fields of vision but also their own represen-tations are comprised in the drone in a real-time simulation. This bodily entanglement with the drone, and its embodied and telepresence effects, is even more pertinent in another genre of the DIY drone scene: drone racing.

Drone racing began in the 2010s, in the US, Germany, Australia, Asian countries, and elsewhere; there are now numerous leagues, associations, and championships for drone racers. In drone races, the drone has to fly along an (indoor or outdoor) track and master obstacles (tunnels, grids, posts, slides, hoops) that require sharp turns, flips, twirls, and loops. You can watch drone races on YouTube, and television networks also broadcast professional races. In addition to the pros, there are amateur and freestyle drone races, where people get together and fly their drones beyond the confines of a racetrack. Drone racing is basically a sport where participants steer drones equipped with cameras. The goal is to complete the racecourse as fast as possible.

Race drones have more blades and rotors to optimize speed and accel-eration, and their cameras are mounted at the top rather than the bot-tom. This camera position enables a first-person view, which means that the drone's perspective merges with that of the pilot. First-person view, also known as FPV or first-person point of view, is the capacity of the viewer to see from a perspective other than their actual location; it thus evokes a situation of telepresence. This is in many ways similar to the perspec-tive of Julius Gustav Neubronner's pigeons (discussed in chapter 1), whose remotely taken images also showed what the pilot (that is, the pigeon) saw. However, unlike the pigeons and their photographs, the drone pilot on the ground sees at the same time and in the same sequence as the drone cam-era. In drone races, the drone camera film is streamed live via radio signal to first-person-view goggles worn by the pilot. The goggles isolate pilots from

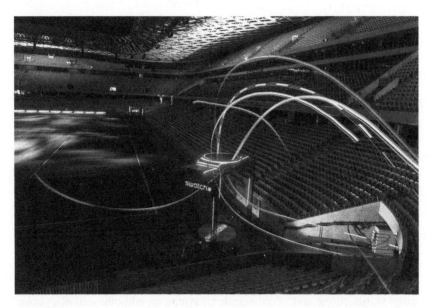

Figure 2.3
Image of a drone at the Drone Racing League Allianz World Championship (2018).

distractions as they watch the material transmitted by the drone cameras. The video that the drone takes can also be seen on large-screen monitors during the race, or watched later in edited versions on YouTube.

Even if you only watch drone racing videos on YouTube (and do not participate in the race live through goggles as a pilot), you quickly become aware that a racing drone does not provide a scopic view from the sky. The impression is more like being inside a video game or on a roller-coaster. The drone camera merges with the eye of the pilot, who in turn steers the drone with their extended camera eye. Pilot and YouTube viewer alike merge with the drone eye and become immersed in its space, movement, and field of vision. There is no longer a sovereign view from above; the drone twirls, makes acrobatic jumps, flips, and ducks as it jets along the track. Most breathtaking of all is the drone's speed (it can reach a hundred kilometers [62 miles] per hour), which can cause vertigo, orientation loss, and motion sickness. Drone racing evokes a sensing based on the body's inner schema: besides conveying the world visually, the drone has a kinesthetic and somatic dimension that affects the whole body.[22] Music plays an important role too, as these drone videos have musical soundtracks—a factor that is also important in one of the artworks I discuss later in this chapter. Indeed, there is a whole genre of music videos shot by drones, which not only have extreme anti-scopic perspectives but also embrace the drone as a synesthetic, embodied, tactile, and acoustic medium.[23]

Thus neither pilots nor viewers of race drones are distanced or detached. Instead, they feel immersed and embodied in the space of the racetrack. A member of a drone race forum describes the experience: "You get to trans-plant your mind, your consciousness, your entire being into a flying object, and you have 100 percent freedom."[24] The mind and the body are extended to the drone, and the drone pilot constitutes a unity with the flying object. It becomes what one drone racer describes as an "out-of-body experience."[25] However, this does not imply that the pilot is disembodied. Rather, the body of the pilot—and not only their eyes—is extended into the machine. As also happens with virtual-reality goggles in computer simulation games, the body's whole sensory system is transposed into a parallel reality. In other words, drone and pilot become an avatar, an embodiment of a person in another reality beyond the pilot's body. The pilot's avatar state is accom-panied by high-adrenaline ecstasy, often enhanced by heavy metal music in edited YouTube videos of drone races. According to the racers themselves, this experience of transplanting the body enables them to sense their bodies in new ways and constitutes a transformative experience of their own iden-tities. Very clearly, the drone works as a device of telepresence, as it immerses the pilot in another space via embodied sensing facilitated by the drone as well as the video graphics array (VGA) goggles. In this way the telepres-ence effect becomes much more intense than on Zoom, for example, since the drone in combination with the VGA goggles affects the whole body and simulates the feeling that the pilot has been elevated into the air.

Drone racers often gather in online communities where they can post their experiences of drone races and upload videos. For example, the FPV Drone Pilots website invites its members to exchange information about drones, meetups, and racer categories.[26] Drone racing is often characterized as a male-dominated sports community—"toys for boys."[27] But in recent years, more and more female drone-flying communities have emerged, and they have established an online presence on websites such as The Drone Girl.[28] These websites organize events where women can learn how to fly drones, attain pilot licenses, and go on collective drone outings. Many female drone pilots consider drone racing to be a form of emancipation, particularly within the context of technology and gender. They often see drone racing as a mind-opening experience, one that has transformed their lives and identities. For example, Zoe Stumbaugh started to fly and race drones after having been ill and physically immobile for two years. She states that drone flying is "like having an out of body experience that you get to control. I liken it to being a monk where you can have an elevated experience and get to leave your body. . . . I like to think of it as being a ballerina in the air. You are quite literally dancing in the air with your drone."[29] The drone enabled the former motorcyclist to race again and experience her body in new ways. It allowed her to dance in the air while sitting in a chair, and even evoked a meditative experience. It triggered a freeing moment in which her body became physiologically extended, weightless, and unbound; in the manner of telepresence, the body could go to other places than its real location.

Drone racing (and dronies) certainly have a community-building dimension. Drone races and dronies are usually uploaded and published on social media platforms (such as Instagram or Facebook), where they can be shared, ranked, and commented on by other viewers. In many ways, the drone race videos and dronies are close relatives of the selfie, the spontaneous digital images we take with our smartphones. Selfies too are always about communities: they can be shared on social networks, and most of those who post them expect recognition, commentary, and endorsement from their peers. As Rose Ehemann has shown, from a psychological point of view, dronies and selfies are mirroring devices that hark back to the identity-shaping function of the silver mirrors of the nineteenth century.[30] The psychoanalyst Jacques Lacan considered that looking into a mirror was crucial for the formation of the child's sense of bodily unity. During the mirror stage, the child sees itself from the outside as an "object" (whether in an actual mirror or in another symbolic representation), thereby shaping a sense of subjectivity.[31] Drone race images and dronies may have a similar effect.[32] To use Peter Sloterdijk's term, they can be "ego-technical devices."[33] They can construct the body and mediate the self in relation to a social environment. The drone races and the dronie create images of the self that are then disseminated to an online community of viewers, who in turn reflect the posted self through their comments and rankings. Beyond such online posts, the drone race and dronie community also exchange technical details, information about equipment, and dronie hot spots online.[34] While there is a strong online component, drone race and dronie communities (and drone lovers' communities more generally) also have a real local dimension. For example, the Airvuz website provides information about drone festivals and competitions where drone pilots can meet physically. Indeed, Airvuz sees itself as a community whose members do more than just post online; they also get the chance to connect in real life. The center of this community is the drone, around which the meetings and events are organized. Like a member of a club, the drone plays the role of connector, mediator, and facilitator of the community. Although these drone communities constitute themselves through social media, newspapers, and television shows, they have a strong real-world aspect.

Thus the community takes the form of an association and a forum— what in German one might call a *Verein* (club). It is a club of enthusiasts and aficionados that revolves around UAVs. This form of community does not seem to align with Jean-Luc Nancy's or Rosi Braidotti's ideas about non-identical communities. Hobby dronists (racers and dronie takers) identify as a community around the drone. Nevertheless, it is interesting that these dronists' communities are not necessarily built on exclusion but rather are defined by a great passion for the drone and for flying drones. In this way, they potentially have a moment of what Nancy would describe as "inoperativity," since they are defined by a common enthusiasm for an object, rather than for a nationalist or exclusive mission. As with the antiquarians of the nineteenth century, it is their love for the objects, not a political agenda, that brings these people together. Of course, the tech industry feeds these communities, and thus one must be aware that the communities are also defined by the drone manufacturers' commercial markets. The hobby drone scene (as well as the professional drone racing scene) does not necessarily represent a critical countercommunity against drone surveillance or the power of tech companies. Dronie social media fora are mostly interested in drone technology, dronie aesthetics, and the embodied forms of vision that dronies enable. Most drone race communities are fairly apolitical, and their primary goals are to have fun and race drones.[35]

Embodied Sensing and Telepresence in Contemporary Drone Art

Artists have been fascinated by technologies of telepresence for a long time—indeed, if we include the robots and seeing machines found in literature, film, and visual art, for centuries. During the 1990s, however, shortly before the boom of the World Wide Web, artists began to engage specifically with telepresence media technology.[36] I am interested in contemporary artists who work with drones as a technology that can evoke effects of telepresence, and how these effects can engender visions of communities.

There have been some literary works that represent drones as telepresence machines. For example, in Tom Hillenbrand's 2014 crime novel *Drohnenland* (*Drone State*), a "mirror space" teleports people to other places, and constructions of numerous invisible drone cameras can simulate faraway situations and spaces in real time, enabling always-on surveillance and eyewitnessing.[37] In William Gibson's 2015 novel *Peripheral*, dronelike androids (the "Peripherals") are controlled by pilots, and the character Flynnie provides security services through first-person-view drones. However, these literary imaginings are not so productive, as they mention drones only marginally and do not engage in depth with the embodied sensorium of the drone. Instead, my focus in this chapter is on contemporary visual artists who explore the constructive and creative practices of amateur drone technology, its potential for embodied and telepresent sensing, and its community-shaping effects. In contrast to DIY dronists, these artists use the technobody of the drone to reflect on the construction of the body and its relationship to machines. In doing so, they provoke reflections about how drone technology and its embodied sensorium can shape imaginaries of communities by revealing the drone's dualistic and dialectical configuration. The artworks simultaneously reflect new, creative, symbiotic communities of drones and humans by simultaneously raising questions about the power and violence of such drone communities.

The artworks in question can all be characterized as installations: they are three-dimensional and site-specific, and they combine diverse media and materials.[38] Thus they share an evocative and performative quality, drawing the audience in as active participants. In this way they transform the audience into active interpreters, and in some instances even cocreators. In what follows, I discuss artworks by Raphaela Vogel and Korakrit Arunanondchai. Although these artists deal with the drone in very different aesthetic, cultural, and political contexts, their works are connected by their common interest in the drone as a medium of embodied sensing and telepresence, and by their suggestion of a community dimension through the projection of aesthetic imaginaries of human–machine convergences.

Raphaela Vogel: Dancing with the Drone

In her expansive installations, Vogel explores the relationship between the body and digital technology.[39] For my purpose in this chapter, the most important of these artworks are her performances and films, in which the main actors are drones and Vogel herself. Her installation *Prophecy* (2016) shows her outdoors, in a blue sky, controlling civilian drones with a remote. Dressed in a swimsuit, she dances and moves in front of the drone cameras. Long white veils cover her body and move in the wind, which is partly created by the drones' propellers. The performance is accompanied by the heavy metal band Soulfly. This installation can take up an entire gallery room, encircling the visitor with two video screens that show the drones flying in front of the artist and a white-sheet backdrop.

Prophecy integrates movement (the artist dancing, the visitor walking around the installation), sound (the loud music, which vibrates in your ears and body), touch (the wind shown in the film), potentially smell (a portable toilet is positioned in-between the two video projection screens), and vision (the drone film). The drones do not film from a scopic, omniscient aerial perspective. Rather, they film from the front, and the cameras are turned slightly upward, into the air. Vogel says the following about her camera technique:

Figure 2.4
Video still from Raphaela Vogel, *Prophecy* (2016). © Raphaela Vogel. Courtesy BQ, Berlin und Raphaela Vogel.

"I do not aim the cameras down, as most do, but up front, so that you always see the device as part of the picture, the propellers look like eyes, it is like a real figure."[40] Similarly to racing drones, Vogel's drones thus offer a first-person view, as the cameras are tilted and aligned with the pilot's eyeline. Thus the drone suggests an embodied sensing that creates an immersive environment from a first-person viewpoint and with a multisensorial quality. The viewer is sucked into this immersive space and senses the dancing artist from the drone's viewpoint. In other words, the drone also displays the dimension of telepresence. It transposes the viewer of the artwork into the perspective of the drone. The drone presents itself as if the audience were the pilot, and it takes us into the middle of the performance. We viewers can slip into the role of the dancing drone; we take over its gaze of desire and are immersed in its sensing field. The array of Vogel's props in front of the film screens, such as the portable toilets, add to the illusion that we are in a simulated space, similarly to the panorama's use of other elements to complete the sensorial experience. In Vogel's drone sensorium, we become dance partners and participants via the sensorium of the machine. Indeed, it is an interesting twist that the audience, rather than the artist, share the drone's viewpoint. Vogel as pilot, filmmaker, and artist does not see what is being filmed while the filming is taking place. There is no immediate feedback, and she shoots the same scene again and again to get the final take. This means that the drones create surreal and distorted images, because it is impossible to take sharp, in-focus images without immediate feedback. Instead of delivering precision (a capacity with which drones are often associated), the drones gain a degree of creative autonomy, taking deranged, kaleidoscopic mirror images.

Far from constituting one all-seeing eye in the sky, the drones in Vogel's installation experiment with multiperspectivism, distortion, and a profusion of visual angles. The drones themselves are always part of the picture, visibly marking the frame of the image (again, similarly to Neubronner's pigeons). Drones and artist interact and engage in an intense dance, generating a sensorial human–machine assemblage. The drones in *Prophecy* immerse the artist, surround her body, touch her with their self-generated winds and sound waves. Like dronies and drone races, Vogel's drone dance performs a somatic and bodily interaction with the drones, which experiments with the drone sensorium as a medium for telepresence.

This assemblage entails a community dimension: a machine–body community in which human and machine are interactive partners. The drones as technical objects are no longer ontologically separated from the subject, but become animated and alive. The object of the drone machine has gained agency. In other words, the idea of a community as an exclusively human assembly is expanded to machines, as the drone stands on the same footing as the human—equal associates in a collective. Unlike drone racer communities, this community is not about connecting a group of people (online or in physical space). Rather, it is based on two agents: the drone and the protagonist/artist in the film. Of course, there are also the implied spectators of the film and the actual visitors to the installation in the museum, but Vogel's artwork shows a community of two. In this interaction, the human agent is no longer alone, in control, or at the center, and instead has to compete with the drone about who steers whom. The drone has become an intelligent partner. Hence the community here is no longer about the identity of a group of subjects. Indeed, this community carries a moment of Nancy's inoperativity insofar as it is a fluid and relational connection of singular agents. One of the agents is nonhuman, and this opens the inoperative community to objects, machines, and drones, expanding the idea of community into a posthuman relational web of human–machine interactions.

In Vogel's installation, however, the drones do not create harmonious pas de deux with the human protagonist/artist; the community displays conflict, tension, and competition. In this way the installation critically reflects on cultural representations of the female body, technology, and the male gaze. The drones in *Prophecy* come across as aggressive media, attacking the artist and attempting to penetrate her body. This aggressivity is intensified by the acoustic space the installation generates, where the drone sounds and heavy metal music craft an intensely heightened affectivity. The drones come to embody mythical creatures from the sky: the whiteness of the drones and the sheetlike veils can be read as references to Pegasus, the divine horse. The female body is exposed as vulnerable, an object of voyeurism, in opposition to the beast–drone machine. In this respect, *Prophecy* shows similarities to Mato Atom's experimental art installation *Seagulls* (2013),[41] which centers on a bikini-clad woman lying on a beach. Drones swarm above her, and one of them descends and seems to penetrate her, turning into a wolf. *Seagulls* not only portrays the violence of the drone through bestial metaphors (birds, swarms, the wolf—see my discussion in chapter 6), but also demonstrates the close connection between drones and the body, and raises questions about how we culturally gender technology. But *Prophecy* deconstructs and ironizes these archaic gender images and power relations (the male gaze, male control of machines and technology) by pushing back against the power of drone technology. Vogel's dance with the drones literally stages a tug-of-war over their control. As the artist mentions in an interview, the drone is at its most creative when it is no longer clear who controls whom: "The tension between control and loss of control is the theme in art."[42]

In *Prophecy*, the human agent has lost its exclusive power to control the drone, because the drone simultaneously also controls the pilot. *Prophecy* suggests a counterprogram to the notion of drones as UAVs. These drones are not remote or distanced. They interact bodily with a human agent; they are immersive and engaging. There is no disembodiment; indeed, the drones make the body present as a sensorial organ, a cultural construction, and a discourse of power. In this light, *Prophecy* gives the idea of the "unmanned" an ironic and literal twist: Vogel's drones literally unman the technology by staging the female body and destabilizing the male fantasy of dominance and control over technology. Somewhat like the Gaga drone, Vogel's artwork queers the drone as it subverts, disorients, and reflects on gender norms and male/female dichotomies.[43] Vogel's art neither engages with military uses of drones nor critiques drone surveillance in domestic contexts. She neither affirms the drone as a new technological fetish nor condemns it as a military technology. Her art performs the dialectics of the drone; the installation stages a queering that invites us to reflect on how we interact culturally with technology. This is also vital in my next example of interactive body drone art.

Korakrit Arunanondchai: Talking to the Drone Spirit Chantri

Working with performance, video, installation, painting, and music, Arunanondchai creates a zone where Buddhist spirituality, animism, memory, data, bodies, and politics come together. Engaging with traditional cultural rituals, the artist creates performances where the audience can interact, participate, and share experiences with each other and with a drone. As I will show, Arunanondchai's drone artworks also include a strong commitment to communities, since the drone becomes an agent in a community with the artist and the participants in the artwork. In 2013, Arunanondchai started his series *With History in a Room Filled with People with Funny Names*. Five versions of this video installation exist, each with a slightly different title and content. In all five versions, however, certain topics and characters remain constant. The recurring central character is a fictional Thai painter, who is depicted in situations that reflect both Western and Eastern cultural beliefs, the natural environment of Thailand, digital technology, tourism, and global capitalism. Another recurring central figure in all five versions is the drone Chantri, which produces footage and acts as a conversation partner for the artist. Chantri's films show images of riverscapes, Bangkok cityscapes, and gatherings of people. The angle of the drone in these films switches: sometimes it takes a first-person view across the landscape, as well as vertical views from above. But Chantri can do more than film: the drone engages in conversation and becomes an interlocutor, a conversation partner, and a medium to connect people. For example, the painter reflects with Chantri about art, the role of artists in Thailand, and what art is: "Chantri, I think I finish my final painting, will you come to see it?"[44] The drone does not answer directly, but transmits images and texts. Chantri speaks in the voice of the artist's mother and shows images of his grandmother, who has dementia. The drone thus embodies someone who is absent: Chantri substitutes the grandmother. In doing so, Chantri emanates the effect of telepresence, as it makes someone who is remote and gone present and embodied. The drone sensorium thus acts as a generator of telepresence. However, it does not absorb the pilot or the viewer into another reality. Rather, the drone itself embodies someone that comes from another place: it animates the presence of the absent grandmother. The drone embodies a "ghost" from the past and makes this person present. Moreover, we do not see the grandmother's body, but we hear her voice and her words. The drone simulates a situation where the grandmother is present, and you can speak to her.

In an interview, Arunanondchai notes, "Chantri is like a filter in the video for the audience. It creates this kind of closeness."[45] The technology of the drone is animated into a spiritual being and conveys an atmosphere of intimacy.[46] For example, the fourth iteration of the video installation was edited according to the breathing sequence of a meditation ritual. Arunanondchai attended these ritual sessions and learned how to enter a state of empathy by focusing on the breath.[47] The drone as a medium in the film sequence is brought right into the meditation ritual, raising questions about whether the drone can breathe too, and whether it can experience empathy as a form of personified technology. The drone enters an intimate community

Figure 2.5
The drone Chantri and the artist. Video still from Korakrit Arunanondchai, *Painting with History in a Room Filled with People with Funny Names 3* (2016). © Korakrit Arunanondchai.

with the artist and his family, and similarly to Vogel's art, this community is based on an affective interaction where machine and human seem to be equal partners. But here the drone also creates larger spiritual communities by connecting groups of people: Chantri plays a central role in creating the flash mobs and dance performances that are key to Arunanondchai's work.

Arunanondchai describes the drone as a spiritual medium and a mythical storyteller: "Yeah. I mean, like the drone in the movie is a kind of spirit. Now we have these powerful prosumer drones, like consumer and producer at the same time. So now everyone can afford to have the eye in the sky. Now everyone is connected."[48] Instead of being a technology of absence, Chantri generates a state of telepresence by taking part in a conversation, making a performance, and bringing people together. Similarly to Vogel's *Prophecy*, in Arunanondchai's work the artist, his body, and his soul form a community with the drone. They are closely connected to each other, and there seems to be an intersubjectivity between them. The drone is thus not so much a prosthesis of extended seeing as it is an embodied, livened conversation partner and community shaper. When a drone seems to be able to engage, show empathy, and create memories, it loses the threatening aura of a killer robot and attains human traits. Chantri is a quasi-object, a hybrid that makes us think about our relationship to technology. In a way, the drone enters into competition with the human, as it seems to have characteristics that have hitherto been exclusive to humans.

Unlike Vogel's work, Arunanondchai's drone community can integrate many people, and his performances often include dozens of participants. As already mentioned, his installations connect groups of people in flash mobs and dance performances. The dance scenes filmed by the drone represent a community in harmony with the drone, and Chantri becomes a spiritual center for the group. There is something archaic and mythic about this flash mob community around the drone. In this context too, the drone artwork displays some of the repercussions of "inoperative communities" as constituted by a certain degree of nondiscursivity and noninstrumentality: the dance scenes convey the impression of a collective in which the individuals have merged into one and been absorbed by the drone.

Certainly, this cultic and mythic view of drone technology and its archaic community vision is problematic, given the duality of drone technology and the common fetishization of drones as a technology with mythical powers, as in the military drone Gorgon Stare (discussed in chapter 1). However, Arunanondchai's art does not celebrate drone technology as an incarnation of hypervisuality, precision, and control. Nor does Chantri simply advertise a fuzzy soulfulness of the machine or rejoice in cultic communities. The conversations and dances with Chantri raise critical political questions about art and global tourism. For example, Chantri's conversations reflect on topics such as the power of art, what is recognized as art, and who is able to become an artist. The people that Chantri films and brings together dance against the instrumentalization of Thailand as a destination for global tourism and sexual exploitation. Arunanondchai notes in an interview, "At this point, capitalism controls all the social media channels, which are our global infrastructure to talk to each other, to represent and understand each other. I was trying to take back some agency through Chantri, giving voice to the in-between space and writing a script for it."[49] Similarly to hobby dronists, Arunanondchai reappropriates drone technology for his own purposes. However, his agenda is much more critical and political. He alienates drone technology from its military and tech industry background, taking it back and transforming it from a cold, distanced, nonhuman machine into an animated one that has a soul and can shape communities. In doing so, he neither affirms nor criticizes the drone industry, but uses, recycles, and reutilizes its products in a creative critique of society and a political intervention against the Western exploitation of Thailand.

In sum, these creative uses of amateur drones by the artists Vogel and Arunanondchai show that such drones engage with the body. Rather than creating distance, they can be interactive, embodied, and immersive. As mentioned, these drones act as what Maximilian Jablonowski has called "presence machines"[50] in the sense that they suggest the sensuous, affective presence of a remote agent. The artist Roy Ascott, who discusses telepresence in his essay "Telenoia," notes, "The technologies of presence are preparing us for connectivity with artificial life, the creation of a cyborg culture. If we are leaving the old, classical, earthly body for another, it is not in order to dematerialize but to inhabit a new corporality, which is almost totally artificial, bionic, prosthetic."[51] Ascott thus sees telepresence as a technology that may offer humans new cognitive-sensory insights, a connectivity with machines, and a transformation of consciousness. Drones' telepresence effects give humans the feeling of being present at a place other than their actual location. This is particularly evident with dronies and drone races, where drones stimulate an externalization of the body into a different time and space. But in the experimental artworks just discussed, the drone can also evoke the effects of telepresence by simulating an animated, soulful, and mindful body.

As shown, these drones as telepresence technology implicate a community dimension: they not only connect individual subjects with technology but can also join together whole groups of people. The drone communities discussed so far in this chapter often accord a creative, playful, even spiritual role to the drone. The power of this telepresence does not turn against the human and subjugate it to the drone machine; rather, the telepresence becomes an agent of sharing, closeness, communication, and commemoration. In the case of hobby dronists, telepresence provides a bonding experience that dronie makers and drone racers use to build their own local communities. In the artworks I have discussed, the drone's telepresence offers a means to build synergetic communities between humans and machines, in which the human is no longer the center point of social, emotional, and intelligent interaction. Although the artworks (especially Vogel's) also allude to the drone's potential violence, they do not question drone technology as such. Rather, the drone becomes a vehicle for reflections on technology and communities, including through its bodily, interactive, and social dimensions. However, when we are dealing with drones—including so-called good drones—it is of the utmost importance that we also show the other side of this technology and demonstrate that this capacity of embodied sensing can also become an instrument of exploitation and human oppression.

Humans as Drones

As early as the 1980s, Marvin Minsky proposed that telepresence technology might be used to deal with accidents in nuclear power plants, or to help humans gain access to secluded and hostile areas. He saw telepresence not only as a technology to handle emergencies, but also as an instrument to maximize profits: "With telepresence one can as easily work from a thousand miles away as from a few feet. Manual labor could easily be done without leaving your home. People could form work clubs. One region of the world could export the specialized skills it has. Anywhere. A laborer in Botswana or India could market his or her abilities in Japan or Antarctica."[52] Minsky's expression "work club" suggests voluntariness and self-organization, euphemizing the realities of outsourced labor. The outsourcing of labor—including virtual labor, such as telemarketing, call centers, and information technology services—is not a voluntary effort on the part of the worker. It is often based on social inequality because it is a cost-cutting measure in respect to salaries, personnel costs, safety, and labor regulations. Although Minsky saw economic benefits in telepresence, it comes as no surprise that his ideas were inspired by Robert Heinlein's dystopian short story "Waldo" (1942), in which a telepresence technology turns humans into its servants.

Today's telepresence technology can use humans as surrogates and turn them into drones—which happens in the imaginaries of engineers and artists alike. But how can a human be seen as a drone? This would certainly be a stretch with regard to the definition of drones as remote-controlled machines. However, if we recognize telepresence as a key signifier of the drone, then the examples I am about to discuss can be seen as cases where humans become drone sensoria.

Let us first discuss an engineer's imaginary. The recent invention of the Human Uber turns a remote-controlled human being into a drone for another human. At first glance, the Human Uber might seem like an art installation, but in fact it was invented by the Japanese engineer Jun

Rekimoto, who developed it in his lab at the University of Tokyo in 2015.[53] According to Rekimoto's website, the technology "uses a real human as a surrogate for another remote user" by giving the surrogate "a mask-shaped display that shows a remote user's live face, and a voice channel [that] transmits a remote user's voice."[54] This mask, known as the ChameleonMask, is an iPad that a human can hold in front of their face. This human is then steered from a remote location by another human, who gives commands and whose face appears on the iPad screen. In this way the controller can interact and be present at conferences, business meetings, or social occasions, without ever having to leave their house. Where Zoom, Skype, and FaceTime video conferencing are often static and inflexible, the Human Uber is more interactive, and its remote agent appears to be more physically present.

The Human Uber takes the drone sensorium to an extreme. It is no longer defined by a single camera eye. Rather, the whole body of a human being, with all its senses, is turned into a remote-controlled sensing device. The drone has become flesh and completely embodied. The Human Uber

Figure 2.6
Jun Rekimoto Lab, Human Uber. © ChameleonMask.

does not fly remotely, but there are many drones that do not fly. Rather, what makes this technobody a drone is the fact that a human is remotely steered in order to enable a telepresence effect for another human. Although the Human Uber technology seems simplistic and undeveloped (how does the human behind the iPad see?), the invention is alarming. The Human Uber demonstrates what can happen when drones as telepresence technology turn into an instrument of power. The drone-human forges relational communities, but the machine and the human no longer exist in a cooperative assemblage. Rather, these communities are defined by inequality, discrimination, and abuse. The Human Uber drone divides people between those that can afford to buy the service and those that have to provide it. The technology becomes a means of exploitation, extracting more value from workers than it gives them.[55] The exploiters improve their profits (for example, by getting multiple tasks done while they use the Human Uber), and the exploited receive less value for their work (just like real Uber drivers). When humans become surrogates within the capitalist system of exploitation, the drone assemblage has clearly lost its attractiveness as a venue for the imagination of new and alternative communities. Drones can oppress people who have less money, and who perhaps come from precarious and vulnerable societal groups, into labor cyborgs. This is also key to my next example, which is taken from the artistic world of film and discusses migrants as a prosthetic cyberworkforce.

The film director Alex Rivera deals with the topic of humans as drones in his movie *Sleep Dealer*. This science fiction film explores relations between Latin America and the US from the perspective of southern Mexico, in a near future when the US/Mexico border has been walled shut. The main character is Memo, an amateur computer hacker from a poor village in southern Mexico. After the death of his father, who was eliminated by a US drone strike on his home village, Memo decides to become a laborer for the company Coyote Tech.

The workers at Coyote Tech use their bodies to steer robots and machines on construction sites in the US. They also work as nannies, apple pickers, and welders. However, these "immigrant" workers are not actually *in* the US: they are a cyberprosthetic workforce. Memo goes to the black market in Tijuana to get nodes implanted into his body so that he will become eligible to work at Coyote Tech. Through these nodes he is able (once plugged in) to remotely steer equipment in the US. The node workers are called

Figure 2.7
Coyote Tech node workers. Still from Alex Rivera, *Sleep Dealer*, Maya Entertainment (2008). © Alex Rivera.

maquiladora (sleep dealers), suggestive of the narcotic state they are in, their dependency on the factory, and the decline of their bodies. They are plugged into machines via the nodes, and all their human senses and physiological muscle power are used to steer the machines remotely. As drones, they overcome spatial and temporal boundaries, performing what Donna Haraway calls "the God trick"[56] of remote surveillance technology. But these workers are not gods in the sky, they are cogs in machines that extract their life force. Memo says, "We call the factory workers 'sleep dealers,' because if you work long enough, you'd collapse."[57] The workers receive no benefits and no insurance; even the cash machine that enables them to send money home to their families exploits them by charging horrendous fees. These machine–human assemblages are not creative imaginaries of machine–human communities. Rather, the assemblages and the workers' bodies are commodities that circulate in the global economy of exploitation.[58] These human drones suggest that global capitalism connects people via remote technology's ability to overcome spatial and temporal boundaries. All the characters in the movie want to be connected—to the big cities of their dreams in the US, to their families and friends at home, to each other. But their fantasy of being connected via technology alienates them into robots, and the connecting nodes destroy their lives. This also reiterates that drones escape strict binaries. There is not always a clear distinction between the scopic (violent, verticalized, military) regime and the nonscopic (embodied, nonviolent) regime. For the human drones in Rivera's film, even though the technology is embodied, networked, and decentralized, it still creates violent and exploitative forms of community. Thus, *Sleep Dealer* effects a critical intervention, and a buffer against excitement about drone cyborgs as creative portals that can redefine the relationship between humans and machines in a posthuman context.

Cyborg Communities

This chapter shows that the drone sensorium has a bodily dimension and can perform modes of embodied nonscopic sensing with telepresence effects. Dronies and drone racing are examples that show the telepresence effect that is inherent in the drone sensorium. Like their nineteenth-century predecessors, they are media that attempt to immerse the viewer in another reality, in a bodily and affective fashion. The artworks I have discussed here experiment with these telepresence effects, either because the drones invite the viewer to be part of a dance performance (in Vogel's installation) or because drones embody other persons (the grandmother drone in Arunanondchai's work).

All of these embodied drone sensing assemblages can also—for better or worse—shape the imaginaries of social communities. These communities can have a local dimension, as we have seen with the hobby dronists. Besides connecting online and through social media, drone racers meet physically and organize community events in ways that are reminiscent of private associations. Thus the drone as a remote surveillance technology constructs not only anonymous collectives but also communities that are reminiscent of (sports) clubs. The drone artworks discussed in this chapter also reveal this community dimension of the embodied drone sensorium. In contrast to hobby dronists, these artists provoke debate about how machine–human communities are culturally and politically envisioned. The artworks by Vogel and Arunanondchai demonstrate the creative potentials that are available for communities when the drone is placed on an equal footing with the human, although they also reveal the dialectics and duality of drone technology. In these artworks, drone–body assemblages generate a type of body that is similar to what Braidotti calls the process of becoming the machine: a body that is thought in symbiotic interdependence with technology. Braidotti states, "The human organism is neither wholly human, nor just an organism. It is an abstract machine, which captures, transforms and produces interconnections."[59] The body is technocultural, a hybrid between the human and the technical. This chapter reveals opportunities to think of the body beyond its binary opposition to the machine, and has discussed how artists use this prosthetic extension of the body as an inspiration to envision new forms of community and social bonding. However, it is equally important to engage with the dialectics of drone cyborg technology. Haraway sees the cyborg as a symptom of the power of global capitalism because it embodies the "informatics of domination."[60] Cyborgs symbolize the fluidity and connectivity of humans and machines via technology, which in turn can also be read as a product of the global flow of capitalism as a network of material and semiotic processes. The Human Uber and Rivera's film are examples not of creative assemblages and communities *with* the drone but of humans oppressed into *becoming* drones. We must remain aware that cyborgs reflect their own origins: they are brainchildren of capitalism and the military.

3   Facial Sensing and Datafied Communities

Perhaps the first association that comes to mind when one thinks about drones is *facelessness*. Facelessness embodies the cruel asymmetry of drone warfare. There are no face-to-face encounters, no opportunity to look back at the perpetrator, who remains invisible and out of harm's way.[1] According to Emanuel Levinas, the absence of the act of mutual seeing—the lack of an intersubjective relation between the "you" and the "me"—erases the preconditions of ethics.[2] The face embodies vulnerability and nakedness; it is a body part that appeals to nonviolence: "The first word of the face is the 'Thou shalt not kill.' It is an order. There is a commandment in the face, as if a master spoke to me."[3] The absence of face-to-face contact has sparked a debate about whether drone warfare is unethical per se.[4]

Contemporary artists who produce works about drone warfare frequently make efforts to restore the victims' faces. The art installation *Not a Bug Splat* is an example. A group of artists—Saks Afridi, Noor Behram, Akash Goel, Assam Khalid, Insyia Syed, Ali Rez and JR's Inside Out Project—and Pakistani villagers unveiled a giant vinyl banner showing the face of a young girl amid a lush green field. According to the installation's organizers, the child had lost her parents and siblings in a drone strike. The enlarged face of the child, with her stern gaze and wide-open eyes, mesmerize the viewer from afar; the grainy black-and-white image recalls military visual material. The grand-scale portrait gives a face to otherwise anonymous killings and breaks the one-sided frame as the girl's eyes look back at the perpetrators.[5] Similarly, director Gavin Hood's *Eye in the Sky* (2015), a film about a military drone mission in Nairobi, displays close-ups of faces: a Kenyan girl who is killed in the attack, and the tearful faces of drone pilots as they press the trigger. The latter images are problematic insofar as weeping pilots suggests that the military system is capable of empathy: we are really sorry, but we have to kill you.

What happens to the defacing power of the drone when it is equipped with facial recognition software? Do drones with facial recognition exist? Is drone footage of faces good enough for facial recognition software to be able to read it? The US military has been investing in the potential for combining drones with artificial intelligence (AI), machine learning, and facial recognition software. Project Maven, an AI program to analyze drone footage, was designed to improve the precision of drone strikes.[6] The machine learning software was supposed to work with intelligence analysts by tagging targets in drone video footage.[7] Some Google employees ultimately boycotted the project, while scholars critically discussed the ethical consequences of such remote algorithmic warfare.[8]

But in domestic contexts too, there exist combinations of drone surveillance and facial recognition. As one can read in consumer advice fora, it is easy to buy a domestic drone with an in-flight facial recognition program.[9] In addition, there have been experiments with drones that react to eye movements.[10] However, there are some technical challenges and legal limitations to the full exploitation of drones' potential with regard to facial recognition. Domestic drones' ability to produce images that are good enough to run facial recognition programs is highly dependent on empirical conditions, such as altitude, distance, and depression (that is, the angle from which the image of the face is taken).[11] Drones' short battery life also imposes limitations.[12] Laws and regulations further limit the use of drones for facial recognition.[13] In the US, for example, police drones face the same rules and restrictions as hobby dronists; for example, the target needs to stay within the pilot's visual range, and there are restrictions regarding altitude and proximity.[14] The police can only use domestic drones for crowd-monitoring, situational awareness, and deterrence: "Police drone users also need to get a special Part 107 waiver or a certificate of authorization from the FAA [Federal Aviation Administration] for usually off-limits actions like operating drones at night or over people."[15]

But let us take a look at how artists are reacting to this new technological feat of the face-reading drone. Through the lens of artworks, this chapter explores the nonmilitary context of drones with facial recognition software, focusing on visual installations and films that reflect how drones and facial recognition are imagined in the domestic sphere.

What does it mean for our views about data processors when a "dumb" flying drone can turn "smart" and learn to recognize faces? What does that in turn imply for the power of this drone sensorium to shape and imagine communities? A drone sensorium with facial recognition no longer senses in a scopic vertical fashion. Rather, as the artworks in this chapter suggest, a drone with facial recognition senses in a networked, datafied, and multidirectional fashion. This kind of sensing also defines the ways in which these artworks envision communities, and it is a dialectical affair. As I will show, some artworks portray the drone's capacity to see faces as a special form of bonding, and reveal affective communities where drones and humans forge an alliance. Other artists—and these are by far the majority—warn us about drones with facial recognition, dismantling their failures, biases, and violence. In these instances the drone tech turns against the human agent and shapes terrifying visions of communities where drones extinguish humans.

Before I demonstrate these tensions, which define these works of drone art, a short excursion into the *aisthêsis* of the technology of facial recognition is in order. This offers us a trajectory through which to explore artistic imaginaries and negotiations of drones' facial recognition programs and their impact on the formation of imaginaries of communities.

Facial Recognition and Cultural Coding

The face is one of our most important features for communication. Our face always works with reciprocity: it only becomes expressive in relation to and in dynamic with others. Facial recognition programs disrupt this reciprocity: machine vision is unable to flirt with us. Facial recognition systems read faces and identities by matching a digital image of the human face against a large database of facial images. These systems are ubiquitous in cellphones, social media platforms (for example, Facebook, Snapchat, and DeepFace), identity verification systems, transportation (for example, flight check-in), and border control. Facial recognition practices are also embedded in the discourse of biometrics, the "science of using biological information for the purposes of identification and verification."[16] Alongside fingerprints, iris scans, and DNA, facial recognition plays a key role in biometrics. A common denominator of facial recognition is the idea that the face is an important index of human identity.[17] Recognizing faces via machines is a complex undertaking that embraces machine vision, human data analysts, and the human development of algorithms, as well as tagging images with metadata and connecting databases with hardware applications. In this way, big-data archives of facial images are created as storage for visual memory, prompting concerns about privacy and data capitalism.[18] Clearview, a company that offers facial recognition software to private businesses and enforcement agencies, has collected images from the Internet (including social media) to form a gigantic database for machine learning.[19]

A key operation in facial recognition processes is that algorithms analyze features from a facial image (such as the nose, eyes, and jaw), and these features are then used to search for other images with matching features.[20] Earlier facial recognition programs from the 1960s were particularly feature-based. Woodrow Wilson Bledsoe, who worked for the US Army Corps during World War II and also for the Central Intelligence Agency, played a key role in the computerization of facial recognition. His work was classified, and many of his research papers were not published until 2013. He developed facial-feature recognition on a graphic tablet (known as a RAND tablet) used by a human operator, who had to extract facial features and coordinates. These facial data about the person (plus their name) were stored in a computerized, searchable database, which thus constituted a "hybrid man–machine system," since the human operator was centrally involved in the data extraction process.[21] As Lila Lee-Morrison has shown, Bledsoe's technology anticipated the eigenface algorithm used for facial recognition in the 1990s.[22] The eigenface algorithm worked by means of principal component analysis, in which a small set of significant features were used to describe the variation among facial images. Principal component analysis transformed faces into sets of essential features, "eigenvectors," which became the components of an initial set of training images.[23] One achieved facial recognition by projecting a new image into the eigenface data set and finding a match with known faces in the database.

Eigenface was the predecessor of many more facial recognition systems. For example, the popular algorithm Fisherfaces has superseded eigenface because it can maximize the separation of classes during the training process.[24] Many of today's newer facial recognition methods work by means of neural networks' deep learning. For example, DeepFace is a facial recognition program created by Facebook. It identifies human faces in digital images by employing neural networks with over 120 million connection weights. It was trained on four million images uploaded by Facebook users.[25]

Are these algorithms—the integral elements of facial recognition software—able to *see*? If one considers seeing to be a physiological process, one that requires living eyes, nerves, and brains, then machine vision does not count as seeing. If one understands seeing as a hermeneutic activity that interprets, opens, and understands the world as a synthetization of knowledge, then algorithms are not able to do that. An algorithm only matches images against a database, and it can only follow an if-then logic. But things are not so easy. Algorithms can see, at least in a metaphorical sense.[26] First of all, the physiological way of seeing might be only a small part of what we understand as seeing. As early as the nineteenth century, Hermann von Helmholtz, who studied the physiology of visual perception, examined the human eye and concluded that it was unable to produce objective representations of the world: "Insofar as the quality of our sensation gives us a message about the peculiarity of the external influence by which it is evoked, it can count as a sign of it, but not as an image."[27] The eye only receives signs and signals from the outside, which the brain then has to construct into a world. In this respect, humans do not see images of the world; rather, the interpretive power of the brain shuttles between world and perception. For Helmholtz, vision is the result of prior cognitive experiences, unconscious inferences that rely on previous cultural experiences of and assumptions about the world (for example, that objects are not normally viewed from below, or that light comes from above). In addition, human sight is dependent on technological inventions (eyeglasses, microscopes) and environments (light conditions, indoors, outdoors). But for algorithms, seeing seems to be different.

It is true that algorithms only match images against preexisting databases, and this can be used as an argument against the idea that algorithms see. But they are intriguingly good at it. DeepFace's neural networks can "recall" the faces of a billion people, a feat beyond human capability. Although this is a statistical mode of seeing, it is a form of perception that cannot be reduced to the merely quantitative. Like human sight, these masses of images generate worlds and create new knowledge; the algorithms can even train themselves. Humans too have to learn how to identify a cat: the human brain needs to be shown a cat (or an image and concept of a cat) before the human can identify it. Unless you are a Platonist, or a rigid follower of Noam Chomsky, you must acknowledge that the human brain, like a computer, must be fed with data to start the cognitive process of learning and identifying objects.

This discussion shows that machine sensing—in this case, facial recognition—makes us reflect not only on our notions of seeing, perception, and sensing but also on our notion of the image. As Trevor Paglen notes, "If we want to understand the invisible world of machine–machine visual culture, we need to unlearn how to see like humans."[28] We have to learn how machines see, understand their visual language, and abandon the idea that seeing is exclusively physiological. Paglen's *Eigenface (Even the Dead Are Not Safe)*, his 2017 project about training data for image recognition algorithms, is an example of such a visual training program for humans.[29] The project shows machine-generated images of faces that to us appear nonfigurative, unreadable, and inaccessible. These automated images must be analyzed in a new way. Here, images no longer speak to human eyes; rather, machine speaks to machine. The images are beyond mimetic representation, "abstractions alien to human perception."[30] Perhaps they should no longer even be called images: Ingrid Hoelzl's notion of the postimage might be more suitable to capture machine vision's collaborative aesthetic. Images are no longer an exclusively human affair, but are collaborative products that have to be thought together with machines.[31] These machinic images do not reproduce reality but generate and construct it.[32]

We must not, however, simply categorize these postimages into a new taxonomy or visual genre. Instead, as Paglen emphasizes, we must understand them as "invisible images [that] are actively watching us, poking and prodding, guiding our movements, inflicting pain and inducing pleasure."[33] These images do something to us; they change how we perceive ourselves and the world. They are not an abstract and objectified form of vision. Some scholars have missed this essential point. If one were to define machinic seeing as a rationalized and calculated operation based on a Cartesian model, then these algorithms would suggest a mathematical truth of vision, and they would therefore appear to be infallible and indeed better than human vision. Although machinic facial recognition is contingent on various software programs, some researchers consider that this biometric technology can achieve the status of objective evidence on the grounds that machines are free from prejudices and stereotypes, and they are not clouded by bias.[34]

Contrary to such arguments, I strongly reject the idea that these technologies perform neutral, abstract sensing processes. Facial recognition is always embedded in an "active, cunning, exercise of power, one ideally suited to molecular police and market operations—one designed to insert its tendrils into ever smaller slices of everyday-life."[35] These technologies exercise power in their individual applications (in border control, for example), and the very fabric of their perception is deeply culturally coded. No matter how "smart" a facial recognition software may be, it still needs a human programmer, user, facilitator, or consumer. And when there is a human agent in the loop, there is a cultural logic at work.

In particular, facial recognition algorithms need a lot of human training. Thus they are not free from human history, contexts, or cultural perspectives. These algorithms are culturally coded as they are developed, fed, and trained by human agents. Hence they are not objective, but relative and constructed by discourses of power. Facial recognition software has to be thought of as a form of assemblage that entwines humans and technologies together. Even when facial recognition programs begin to learn and program themselves, the cultural context is still present. Simone Browne has demonstrated that facial recognition technologies are deeply racialized and biased.[36] Making a connection between branding as an early form of biometrics and current biometric technologies, she shows how these technologies misrecognize and misrepresent Black people. Facial recognition programs privilege whiteness and lightness, mismatch the gender of African Americans, and use normative racial definitions.[37] Similarly, Joy Buolamwini and Timnit Gebru's study "Gender Shades"[38] demonstrates that facial recognition programs work best on light-skinned males, as highlighted in the popular Netflix documentary *Coded Bias* (2020) about MIT Media Lab researcher Buolamwini's work. Artworks about facial recognition and drone use address precisely this cultural embeddedness and political coding of surveillance technologies.

Artworks on Facial Recognition and Drones

Artists have engaged with the topic and technology of drones with facial recognition within both mainstream culture and experimental visual art. Some of their artworks display a strong belief in the power of drone technology to use facial recognition. In popular culture, drones seem to be able to move around and recognize faces instantly in real time, from any angle or distance, without technical difficulties. There are no battery problems or blurry pictures. Think of the Hollywood film *Minority Report* (2002) and its crawling spiderlike drones, which can intrude easily through any niche, slither up your body, and scan your retina. They are so accurate that the only way to outsmart them is to get a new pair of physical eyes, as the protagonist John Anderton, simultaneously the head and the target of his own precrime unit, ends up doing. In "Hated in the Nation," a 2016 episode of the Netflix series *Black Mirror* (discussed in depth in chapter 6), swarms of artificial killer drone bees equipped with facial recognition can destroy their targets in the most gruesome ways. Both "Hated in the Nation" and *Minority Report* follow sci-fi tradition in speculating about the future in light of new technologies. By means of fictive conjecture, they make us reflect on the societal ramifications of such technologies. These aesthetic imaginaries conjure scenarios regarding the technologies' sensorial features and how they might affect, shape, and organize human communities. Although "only" imagined by artists, these imaginaries speculate about the potentials of the sensorium of a drone that has routinized facial recognition. *Minority Report* posits the societal acceptance of the drone eye scan and gives insights into the drone's motility, its datafied networking ability, and its "intelligence" in recognizing faces and predicting crime. It is precisely the aesthetic faculty of exaggeration, the feverish and nightmarish fantasies of what technology might do, that makes us think about the power of these technologies and their effects on communities.

But in this chapter I am interested in artworks that are more avant-garde, independent, and experimental, because these specifically explore the community-shaping aspects of the technology in question. Contemporary artists and activists have become interested in the new biometric technologies of facial recognition and their cultural logic of exclusion. Many works that engage with facial recognition articulate countersurveillance measures, such as blocking, hiding, and obscuring visibility. For example, Zach Blas makes distorted, colorful, soft ice–shaped masks that simultaneously disfigure and hide the face.[39] The amorphous and opaque masks are generated from biometric facial data, and their distortion marks the obfuscation of race and gender found in so much facial recognition software. Most important are the artists that experiment with facial recognition surveillance technologies and combine them with the drone's sensorial possibilities, such as its aerial mobility, embodied viewing, and remote sensing. Artworks that engage with facial recognition and drone sensing represent an interesting technosensorial mix that in turn evokes a plethora of community visions, as the rest of this chapter explores.

Apophenia, Drones, and Facial Recognition in the Sky

Cloud Face (2012) is a visual installation by the Seoul-based artist duo Shin-seungback Kimyonghun, which consists of engineer Shin Seung Back and artist Kim Yong Hun. Their photos show clouds in the sky that look like fluffy faces with eyes, noses, and mouths. Some seem to smile; others seem to shout something to us. *Cloud Face* is about facial recognition, but it also plays with a drone gaze. The camera's gaze reiterates the gaze of the drone by scanning, searching, and evaluating the environment for biometric data. This simulated drone gaze scans faces, but it does so in reverse, as it does not look down from the sky but up into the air. In doing so, it no longer adheres to the scopic regime of verticality but instead displays a form of volumetric sensing, which suggests a three-dimensionality of space (discussed in chapter 5). This aesthetic mimicking of the volumetric drone gaze is capable of facial recognition. During the making of this visual artwork, a static camera that was pointed into the sky took images of the cloud formations. A facial recognition program then scanned and processed the filmic material. Any time the program saw a facelike configuration, it fixed it as a cloud face. These cloud faces were then presented as an installation where squares of facial images were assembled into a pattern of framed pictures. The installation's symmetrical framing conveys a rationalized counterpoint to the amorphous, magical images of clouds, an effect that reflects the hallucinatory aspect of pattern recognition.

How should we interpret the cloud faces the machinic gaze detected? This phenomenon of seeing things that are not there is known as *apophenia*, the perception of meaningful patterns among unrelated things. For example, children believe they can see a man in the moon, and gamblers perceive patterns and winning streaks against the odds. In such instances, we make sense out of seemingly empty data (such as white noise or static) and find patterns of meaning in them. The psychiatrist Klaus Conrad saw this as an early stage of schizophrenia's delusional thinking and overinterpretation.[40] However, today apophenia is regarded as a stage of human cognitive development.[41]

Figure 3.1
Shinseungback Kimyonghun, *Cloud Face* (2012). © Shinseungback Kimyonghun.

Furthermore, apophenia no longer seems to be a human quality: machines are disposed toward it too. The artist, filmmaker, and scholar Hito Steyerl notes that apophenia occurs in conjunction with data analysts and machine learning programs: "Apophenia is about drawing connections and conclusions from sources with no direct connection other than their indissoluble perceptual simultaneity, as Benjamin Bratton recently argued. One has to assume that sometimes, analysts also use apophenia."[42] Steyerl further states that machine vision (algorithms, and machine vision learning programs, such as facial recognition) can show symptoms of apophenia. Together, the analysts and data processors can become delusional and hallucinate.[43] The Google machine-learning program DeepDream reveals this form of "magical thinking."[44] This program uses convolutional neural networks to detect patterns in images and then deliberately overprocesses them. In doing so, it can generate images that have a hallucinogenic quality. For example, there is an image of a plate of spaghetti that you would never want to eat, since eyes and a dog's face are emerging out of the tomato sauce.

Instead of merely identifying an image, DeepDream enhances what it sees and overinterprets certain features. When the image is back-propagated into the software, it displays surreal and psychedelic features.

In *Cloud Face*, the photos of the clouds in the sky stage the hallucinatory potential of machine vision. The moment of apophenia lies in the machine's ability to see the faces. Gazing into the sky and finding faces in the clouds is a children's pastime, but now it is a machine that sees faces in a "sea of data."[45] Facial recognition scientists have programmed software to see faces within "dirty data," that is, in unspecified data, with the result that it (like us) can also recognize faces that consist of minimalistic indicators ("dot, dot, comma dash—smiley face in a flash"). In short, these machines see faces where there are no faces.

What do the fluffy faces that the facial recognition program finds in the sky tell us about machine vision, and about drones with facial recognition software? What imaginaries of communities do the artworks conjure via this facial sensing technology? In my view, *Cloud Face* does not evoke a cold machine gaze that misconstrues patterns and leads to fatal consequences. Rather, the amorphous faces in the sky evoke fond memories of a children's game. The clouds that the imaginary drone has scanned for us inspire our poetic sensibility: they look beautiful, peaceful, even sublime. They recall Thomas Scott Baldwin's ecstatic descriptions of the clouds and atmosphere that he saw during his tour in a hot-air balloon (chapter 4), or Luke Howard's influential nomenclature of cloud types in 1802. Besides names for clouds, Howard's scientific prose contains poetic and romantic descriptions of these celestial and atmospheric phenomena.[46]

In Shinseungback Kimyonghun's art, the clouds also gain an aesthetic quality of serenity and beauty. They suggest an animistic perception of nature as something that can look back at us. The cloud fluff literally looks at us. When Walter Benjamin traced the remnants of auratic experience in modern society, he mentioned that it emerges in the moment when an inanimate object opens its eyes, communicates with us, and returns our gaze: "Experience of the aura thus rests on the transposition of a response common in human relationships to the relationship between the inanimate or natural object and man. The person we look at, or who feels he is being looked at, looks at us in turn. To perceive the aura of an object we look at means to invest it with the ability to look at us in return."[47] The return gaze of inanimate objects suggests a form of experience that animates the object and renaturalizes the human. The art installation shows that machines have a prediscursive way of sensing. Nature as the representation of the sky communicates in these *Cloud Face* images: it smiles, speaks, looks, flirts, gets into a close dialogue with the viewer. The simulated drone gaze into the sky brings us to this vivification of the clouds: it is the drone that makes the faces come alive. Very unlike artworks that cast the drone as a medium that defaces and brings death, in *Cloud Face* the simulated drone sensorium has an animistic power.

According to Benjamin, nature's return gaze is a way of relating that embraces the subject with the object. There is no longer a distance between these two ontological spheres; rather, they come closer, in a prerational and nondiscursive form of community. This romantic community is constituted by a mimetic faculty, by the ability to make oneself similar to nature, to become nature without dominating or instrumentalizing it. This is not a community vision that consists of a certain group of people. Instead, it is constituted by a certain way of relating to each other, a nonhierarchical connecting between humans and objects, a community that is more of a communion between ontological layers. These are the layers of being a human subject, being a technological object, and being nature. In *Cloud Face* these layers merge and are no longer clearly discernible. This type of communion (as a form of community) does not suggest that unitary totality in which there are no differences. Rather, recalling Jean-Luc Nancy, the community is thought of as a relational construct where different ontological layers can connect with each other by neither creating hierarchies nor flattening their differences; it is more about connecting them in relations and momentary formations.

Cloud Face shows that machine vision can imitate this mimetic faculty to animate nature, and in doing so it summons affective energy to embrace the human in an experience of union with nature. The faces diminish the distance between nature and us. They become poetic symbols of transcendence that open up a provisional frame within which we can rethink the ontological differences between human, nature, and machines. It is notable that the entwining of the modes of being nature and being human is enabled via technology, namely, a facial recognition program that plays with a drone gaze. This technology generates these relational assemblages, which question a human-centered understanding of community. In other words, *Cloud Face*, with its inverted drone optics and facial sensing, suggests that humans are no longer the only ones that are prone to fantasy, creativity, and poesis. Machines can do all of that too, and we do not necessarily have to find it threatening. In many ways, this art installation illustrates Rosi Braidotti's contention that the blurring of boundaries between humans and others (objects, nature) does not necessarily pose a threat or a loss. Rather, it creates a creative trajectory to imagine new forms of social bonding and community building.[48]

However, my romanticizing perspective on *Cloud Face* is only one layer of interpretation. *Cloud Face* also shows how wrong machine vision can be. The cloud faces demonstrate machines' tendency to misread faces and to mark the sky as a territory of atmospheric surveillance and control. Although we are enticed by their associations with childhood memories, these images also suggest that machine vision is fallible. Machines can hallucinate targets, fabricating faces in order to deface. And it is precisely the aesthetic discourse that can make these vulnerabilities and weaknesses of machine vision and drone technology visible. Images of faces produced by drones with facial recognition software can have terrible consequences: they can land innocent people in jail; they can tag and create communities of wrongfully assembled targets. Moreover, the artwork *Cloud Face* hinges on these dangerous sides of machine vision technology: under the surface of these fluffy faces lies a vision of community based on exclusion and social sorting. There is the mimetic and animistic aspect of *Cloud Face*, but the images also demonstrate the great potential for machine vision to go wrong. In doing so, the technology then helps to wrongfully arrest people, to sort them wrongfully into communities of criminals and suspects. This dangerous side of machine vision is the theme of my next artistic example.

Slaughterbots: Domestic Killer Drones with Facial Recognition

Speaking about *Slaughterbots*, CNN's Matt McFarland remarked that "perhaps the most nightmarish, dystopian film of 2017 didn't come from Hollywood."[49] Indeed, this seven-minute video about domestic drones that kill senators and students is a powerful warning of how drone technology equipped with facial recognition might go completely wrong. Directed by Stewart Sugg and narrated by Stuart Russell, professor of computer science at the University of California, Berkeley, the video was released on YouTube and went viral, with over two million views.[50] It merges the genres of fiction and documentary, projecting a fictional scenario of the mass killing of citizens by terrorists using drones. The film presents prosumer drones that not only can shoot but also have cameras with facial recognition software installed. This combination of facial recognition and drones makes it possible to identify, chase, and execute targets in real time. In this way, the sensoria of these drones enable visions of communities that are based on technological violence. The drones in the film terrorize a community of young students and eliminate those that might be critical of the state and the government. The technology of the drone becomes a tool to shape a totalitarian community where violence against individuals is not controlled by the state but executed by terrorists. Although no drones with facial recognition sensors and munitions currently exist in the form in which the film presents them, the video suggests that they are likely to exist in the future. The scenario looks authentic, thanks to reality effects, such as the integration of clips from CNN and British newscasts, and a scholarly final commentary from Russell about the dangers of autonomous lethal weapons. Russell was a key figure in the production of the film, which was made in collaboration with the Future Life Institute, with which he is also associated. The film was screened at the United Nations Geneva Convention in 2017.[51]

Slaughterbots shows how small, domestic-looking microdrones loaded with explosives might wipe out citizens. To be capable of this, the drones are endowed with a powerful technological sensorium, which is praised at the beginning of the film by the sleek CEO of the security firm Stratenergetics.com. Standing in an auditorium, he demonstrates the beauty of the drone to his audience, showing them a slide about the drone's sensorial features: "It has camera sensors, and just like your phones it has facial recognition. . . . It has stochastic motion as an anti-sniper feature."[52] The CEO onstage interacts with the palm-sized drone, playing catch with it to show how it can evade his advances, circle, and follow him. We also see the CEO through the drone, the metadata on the margin of its field of vision, and the hazard sign when the CEO tries to catch it. We see the facial recognition software at work as the film takes the drone's perspective and provides us with all the data about its potential targets. The drone not only takes images of faces but also analyzes these images, searching and matching them in databases in real time. It is this capacity for instant facial recognition that makes the drones so deadly. Each drone is loaded with an explosive that is set to shoot directly into the target's brain.

Despite its lethal load, the CEO evinces an almost tender relationship with the drone as it lands softly in his hand like a butterfly. He engages with it as if it were one of his own body parts, bringing to mind the idea of the drone as cyborg. The CEO and the drone constitute a deadly dance-like assemblage of human and machine: not a beautiful pax de deux, but a danse macabre. As the film unfolds, the drones go on to kill eleven US senators and over eight thousand students in Edinburgh. They are steered by terrorists who have somehow gotten their hands on the technology (it is never made clear who the terrorists are). While the motive for the senators' assassination remains unsolved, the students are apparently targeted because they have shared political activist and anticorruption sites on Facebook. The microdrones are programmed to target specific individuals, as becomes tragically evident when a young student is tracked down by the slaughterbots while on a FaceTime call to his mother. FaceTime ends; we see the blackened screen and the not-connected sign on the mother's phone; the student is dead. The drones and their kill mission are inescapable, precisely as the CEO boasted: "They cannot be stopped."[53]

The slaughterbots are not merely cameras in the sky that watch people. Their sensing modes are datafied and networked as they analyze, track, and search for biometrics. Thus the human face turns out to be the most vulnerable human feature. It is notable that faces are omnipresent in the film. There are the faces of the students being attacked, the faces of the wounded, the faces of the news crews, the face of the terrified mother who watches her son die. These faces build a mosaic of emotions—fear, terror, sadness, empathy—and reveal human vulnerability. To echo the words of Irma van der Ploeg, the bodies of the students become a matrix of information and thus the main source of data for biometric surveillance.[54]

Slaughterbots shows how surveillance has changed from panoptical practices to always-on and fluid ones. These drones exemplify what Zygmunt Bauman and David Lyon call *liquid technology*: no longer perpendicularly ordered, but acentral, flexible, and fluid.[55] The slaughterbots can sense from all kinds of spatial angles, and they can come inside buildings into the crannies of classroom furniture. They do not give up until they have achieved their deadly mission. The drone's facial recognition feature illustrates that surveillance has not only become fluid but can also use the body's biometric data—its physical features such as skin, eyes, body shape, age, sex, and size—as data sources.

Lyon has analyzed this change in surveillance practices with regard to biometric data, and he argues that surveillance is no longer panoptically organized but rather is decentered and embedded in the everyday. In this way surveillance has also acquired a voluntary quality. Facial recognition, DNA, and other biometric data found on social media are examples of voluntary surveillance, which is no longer exclusively performed behind prison walls or on convicted criminals. Indeed, *voluntary* is a somewhat problematic term. Perhaps not everybody is aware that when you "voluntarily" post images on Facebook, they are used for surveillance purposes or for the development of facial recognition programs. In *Slaughterbots*, the students have used Facebook to organize politically, and their posts make them targets of biometric surveillance. This biometric surveillance executes social sorting by categorizing people into specific social classes.[56] Referring to new forms of biometric surveillance, Lyon outlines how surveillance as social sorting is a powerful means of reinforcing long-term social differences.[57] In other words, surveillance technologies such as facial recognition can construct communities based on processes of classification and categorization whose criteria are essentialist and binary. Facial recognition algorithms are good at seeing the world through binary oppositions, such as female/male, Black/white, or rich/poor. They do not do nuance: gray areas and in-between spaces are not their strong suit.

Facial recognition relies on social classification systems that date from the nineteenth century and often have their origins in eugenics, binary theorizations of gender, and essentialist conceptions of race.[58] Facial recognition and its normative assumptions about identity, race, and gender can be traced back to the histories of physiognomy, phrenology, and eugenics. The most prominent figure in this history is the nineteenth-century polymath Francis Galton, who was interested in statistics and anthropometrics. His use of composite photographs to create a taxonomy of "criminal faces" has attracted attention in recent work on facial recognition.[59] Galton invented the technique of superimposing photographs of human faces, using multiple exposures of the same photographic plate in order to attain an average face (by virtue of the statistical mean). In other words, he used photographs to construct "natural types" of races and criminals. Through his attempts to classify humans and create human taxonomies, Galton built a key trajectory for determinist theories about race and gender in the twentieth century. Although today's facial recognition technology is geared toward identifying individuals and their particularities, researchers and activists have shown that it is based on typologies and essentialist notions of identity that can be traced back to these nineteenth-century anthropometric theories. Surveillance technologies such as facial recognition thus execute a form of biopolitics that reflects power relations and policies that have defined modern societies for centuries.

In *Slaughterbots*, the drones suggest an essentialist model of community, eliminating subjects that are supposed not to belong. Showing an animated image of a brain being destroyed by a killer drone, the CEO emphasizes that drones are there to get the "bad guys," and that they can differentiate between the evil and the good: "Now trust me, these were all bad guys."[60] Here again, the drone executes a form of biopolitical power, shaping communities of bad guys versus good guys, those who should be protected and those whose lives should be terminated. Indeed, the corporate ideology of the security firm (represented by the CEO) uses the same rhetoric as the terrorists, who also target specific people (in this case, activists) that constitute a threat from their point of view.

The slaughterbot drone's biopolitical power still depends on humans, however. The human is still in the loop as an active agent steering these machines. The slaughterbots are not a form of self-organizing AI, and the algorithms do not take over complete control from the humans. Although the film is about autonomous weapons, these drones do not have autonomy or consciousness. This drone swarm does not embody the self-sufficient intelligence so often fantasized in Hollywood movies—for example, the fictional machine-learning program Skynet, which is based on artificial neural networks and plays a key role in the *Terminator* movies.[61] Skynet develops autonomy and threatens to eradicate the whole of humanity in a nuclear attack. The slaughterbots can fly by themselves and are on a kill mission, but they did not program themselves to do so: they have been programmed by a human intelligence, even if it is unclear whose.

In a response to *Slaughterbots*, Paul Scharre complained that the film presented an unrealistic and fearmongering vision of autonomous weapons.[62] He argued that the government would never produce such weapons in such quantity, that one would not be defenseless against this type of drone, and that is unlikely that terrorists would obtain such weapons. Russell counterargued by emphasizing the likelihood of the scenario shown in the film. He stated that *Slaughterbots* was not science fiction, and that he and his team had hoped to give the public a warning about technological developments.[63] He referred to the power of the film *The Day After* (1983), which sparked intense political and public debate about nuclear weapons. It is not so important to me whether *Slaughterbots* is a realistic or unrealistic technological scenario. Instead, I am concerned to show that art—in this case, a documentary-style fictional video—has the power to shape reflections on, and even public debate about, the biopolitical power of drone technology. The film proves that art can illustrate, exaggerate, experiment, predict, and warn about the sensorium of drone technology in combination with facial recognition—a sensorium that has left the scopic regime of vertical surveillance far behind. As this book is about civilian drones, I am specifically interested in how aesthetic works imagine facial recognition and drones in the domestic sphere (such as the schools shown in *Slaughterbots*).

The following discussion of some specific artworks demonstrates that facial recognition and drone technology have crept into the realm of everyday life. It is important to remember that the everyday does not equate with nonviolence. For many people, the everyday is a military warzone, and the domestic realm is militarized. Nevertheless, I am interested in drones with facial recognition whose primary application is not defined by military missions but which are integrated into the civilian sphere.

Adam Harvey: *CV Dazzle* and *Stealth Wear*

In response to the increasing presence of drones, the artist Adam Harvey has engaged with the question of how to make oneself invisible. His antidrone fashion project *Stealth Wear* (2012) comprises a set of garments designed to ward off the sensorium of drones and make the wearer almost invisible. A predecessor to *Stealth Wear* was Harvey's 2010 project *CV Dazzle*, which dealt with facial recognition and played an important role in the development of *Stealth Wear*. *CV Dazzle* comprises portraits of people's faces wearing a specific makeup style that disrupts facial recognition software. On a corresponding website, the artist posts tips on makeup and hairstyles to trick facial recognition programs (strong contrasts on the skin, with white patterns on black skin and black patterns on white). According to Harvey, it is a good idea to obscure the nose bridge and the eyes, and to cover up the symmetry of the face. His photos portray the results of these style tips. In *Look N° 5 (b)*, for example, the woman's face is half concealed by a long blue and black ponytail, the eyes are covered, and one cheek bears a rectangular black mark.

Harvey describes *CV Dazzle* as a "type of camouflage from computer vision. It uses bold patterning to break apart the expected features targeted by computer vision algorithms."[64] In this way it tries to find the weaknesses in a facial recognition algorithm named Viola-Jones Haar Cascade.[65] Camouflage, however, might not be quite the right expression for Harvey's endeavors, as the face does not merge with the background or mimic its environment. Rather, the faces dresses up, becomes carnivalesque, and dazzles the viewer. The word *dazzling* also refers to the practice during World War I of painting battleships with hyperbolic shapes and patterns (often stripes) to confuse the enemy.[66] Harvey uses a similar technique to create an "antiface" that becomes invisible by becoming a visual spectacle. Instead of hiding the face, his photos get "in the face" of the algorithm.

CV Dazzle was the basis for *Stealth Wear*, Harvey's work on drones: "Building off previous work with *CV Dazzle* camouflage from face detection, *Stealth Wear* continues to explore the aesthetics of privacy and the potential for fashion to challenge authoritarian surveillance technologies."[67] Harvey designed *Stealth Wear* as an antidrone fashion line, with hoodies, hijabs, and burqas that obscure the body. The clothes suggest that the drone is not just a camera in the sky that takes pictures: the use of heat-resistant fabric indicates that the drone has a more-than-optical sensorium, a sensorium that is not only scopic but also datafied and networked. The artwork assumes a drone that detects body temperatures via infrared cameras. The fabric is covered with metal-plated fibers that deflect and diffuse the thermal radiation emitted by a body under the observation of a long-wave infrared camera. Military drones are frequently equipped with such thermal cameras, which sense body heat and identify targets accordingly.[68] Further, Harvey's clothes seem to shield the wearer against facial recognition software. Not only do the clothes hide the face but the face is also adorned with heavy makeup reminiscent of *CV Dazzle*.

On the one hand, *Stealth Wear* has military connotations. The title itself alludes to a stealth fighter, that is, a jet that can fly "under the radar" by using technologies that reduce the chances of detection. The burqa also suggests military associations by connecting the antisurveillance wear to drone operations in Muslim countries. The burqa in Harvey's fashion line raises questions about a Western gaze and its perception of women and culture. When asked about the role of the burqa, Harvey responded, "The 'anti-drone' burqa piece is full of conflict."[69] This response acknowledges Harvey's own Westernized perspective, which resonates in *Stealth Wear* and situates the burqa as a complex cultural phenomenon. Veiling has many different practices and purposes.[70]

On the other hand, I think that there is another layer to *Stealth Wear* that goes beyond the military context. The word *stealth* can also refer to a person who has passed into their desired sex or gender without revealing their gender history.[71] Harvey's *Stealth Wear* shields users not only from military drones but also from the gender biases of facial recognition software. These clothes make one invisible to drone sensing and also to the facial recognition software that perceives gender only according to stable binaries. Shoshana Magnet has shown that facial recognition programs produce errors in detecting the identities of transgendered people: "In general people who cannot easily be categorized as either men or women are interpreted as biometric system failures."[72] This categorization in turn affects the imaginaries of communities: "However, as these technologies are specifically deployed to identify suspect bodies, the impact of technological failure manifests itself most consistently in othered communities."[73] These "othered" communities are marginalized and made outsiders. In other words, *Stealth Wear* also plays with machine vision biases in respect to gender, and the fashion context of Harvey's work enables users to cross genders, play roles, and take on other identities. Through dressing up and dazzling, this fashion line responds to surveillance technologies by taking a hyperbolic, carnivalesque, and satirical perspective. Thus *Stealth Wear* pulls the topic of facial recognition and drones into the realm of the domestic, the everyday, and immediate lifeworlds. How should you dress every day in a machine-readable world? Harvey's work shows how surveillance technologies have penetrated our domestic and everyday communities, particularly through the commodification and capitalist circulation of garments, and his art emphasizes the embeddedness of surveillance in our everyday world. *Stealth Wear* turns garments into aestheticized commodities and the art viewer into a consumer.

There are parallels to be drawn between Harvey's *CV Dazzle* and *Stealth Wear* projects and Hito Steyerl's film *HOW NOT TO BE SEEN: A Fucking Didactic Educational.MOV File* (2013). Steyerl's film provides "lessons" on how to evade and obstruct facial recognition programs. Svea Braeunert has shown that Steyerl's images "hinge on exploiting fundamental differences between human perception and machine vision. The forms of invisibility they practice are entirely visible to the human eye, but invisible to the eye of the machine, because they confuse its script of automated recognition."[74] Like Harvey, Steyerl dazzles the algorithm by dressing up for it in order to confuse it. She also gives hands-on advice regarding what one can do to remain invisible. Her pedagogical gestures are further ironized through the artistic composition of her film and its title. Although there is a good amount of humor in her work, this should not overshadow the seriousness of the installation, which alludes to people who are "disappeared" and eliminated by autocratic governments.[75] Steyerl explicitly connects her ideas about dazzling machine vision and facial recognition software to the sensing of drones. She explains that drones were the inspiration for the film: "It started with a real story that I was told about how rebels avoid being detected by drones. The drone sees movement and body heat. So these people would cover themselves with a reflective plastic sheet and douse themselves with water to bring down their body temperature. The paradox, of course, is that a landscape littered with bright plastic-sheet monochromes would be plainly visible to any human eye—but invisible to the drone's computers."[76] Similarly to Harvey's work, Steyerl's art thus has a military dimension, as it originates from an observation connected to drone warfare. At the same time, Steyerl reveals the dimension of the everyday and the domestic by setting her anti-surveillance measures in cities and urban environments.

In their visual art, both Harvey and Steyerl engage with the paradox of surveillance: surveillance technology strives for total visibility and at the same time seeks to remain invisible. Their response to this tension is that one should become invisible, obscure, and opaque, even if that means dressing up. This process of becoming invisible by dazzling reacts to the drone as a synesthetic sensorium: makeup and garments counteract the multisensoriality of the drone, which can read faces through datafied sensing, feel heat by detecting body temperatures, and interpret kinetic movements. Harvey's art in particular performs an act of sousveillance. In aesthetic discourses, sousveillance has become an expression about surveillance. It refers to going off-grid, disappearing, and making oneself invisible to surveillance technologies.[77] Sousveillance is an act of resistance in which the individual strikes back by using, experimenting with, and recycling the very monitoring devices that are used for surveillance. Harvey's sousveillance art does not suggest that we should hide or do nothing. Rather, his art seeks to actively shape communities against surveillance. By giving hands-on, practical advice about how to become invisible, his artwork calls on other people to do the same. In this respect, his work parallels James Bridle's *Drone Shadow* (2017), which presented the public with an easy-to-download manual showing how to draw drones on pavements and streets, to raise awareness about military drone strikes. In Harvey's art projects too, there is an activist, somewhat local community-building aspect. People can try out this type of makeup (indeed, many Facebook users have done so, in their very own styles), or they can buy a *Stealth Wear* garment (if they can afford one) or make something similar themselves. Harvey's artworks seek to build a community who want to go off-grid together and playfully make themselves disappear in the age of total visibility. *Stealth Wear* issues a warning about a community that has routinized a 24/7 total surveillance practice in which technologies such as drones, facial recognition, and AI have gained the upper hand. In contrast to *Cloud Face*, this community is no longer defined by a creative assemblage between sensorial technologies and the human. Rather, the entwinedness of the human and the machine, and of the intelligent and sensorial capacities of the drone with facial recognition, are a source of danger, domination, and violence.

Datafied Communities

Cloud Face, Slaughterbots, CV Dazzle, and *Stealth Wear* have demonstrated the speculative power of art that visualizes how facial recognition drones might shape our future communities. Although drones with facial recognition are still in their infancy, as many technical and legal challenges remain, the aesthetic realm teems with scenarios about this technology. The artistic examples discussed here all experiment with the synesthetic sensorium of the drone and reveal its more-than-optical capacities. Although visual sensing is still very present in these works, all of them also work with datafied and algorithmic sensing (facial recognition). The drone's technosensual capacities, displayed in these aesthetic works, are also a matrix for imaginaries of communities. *Cloud Face* suggests an affective alliance between the human, machine vision, and the drone. This community scenario dissolves the opposition between machinic and human vision in order to experiment with new forms of bonding and being in the world. *Cloud Face* suggests a community based on nonhierarchical relations between machines, nature, and humans. Thus its vison of community may be compatible with the thinking of Nancy and Braidotti, who have attempted to describe communities beyond opposition, identity, and binaries (as discussed in chapter 1).

But not all of my examples in this chapter follow this creative take on the datafied sensorium of the drone. *Cloud Face* also reveals the vulnerability and danger of facial recognition technology. *Slaughterbots* warns that drones with facial recognition technology might become instruments of terror and violence. The film suggests that even if the drone no longer works in scopic ways, it can execute power and violence nonetheless. It is no longer a form of verticalized power, but one that is networked, datafied, and decentralized. *Slaughterbots* seeks to reveal this form of power, intervene in political and public discourses, and change society with and through art in a concrete and operative manner. This interventionism is also at stake in Harvey's art. His artworks push back against potential surveillance by drones with facial recognition, using the gesture of sousveillance: the ironic performance of going off-grid and disappearing. In all these examples, diverse as they are, the artworks have shown that machinic vision does not represent reality but has generative power: it constructs our lifeworlds and communities.

III   The Earth

4 Flattened Sensing and Planetary Communities

"Seeing the world from above doesn't just flatten things. It sharpens them," says the actor playing the drone pilot in Omer Fast's fictional short *5000 Feet Is the Best*.[1] Flattening is an iconographic feature of the images that drones produce while surveilling the earth from above, which show the world as an abstract surface without depth. Research on the scopic regime of military drones often interprets these images as products of an operative form of sensing that dehumanizes, controls, and masters the world.[2] As Fast's drone pilot says, the flattened drone view sharpens things: it acts like a violent blade that cuts and kills. But flattening can also have a different effect. In contrast to the military drone, the amateur drone's flattening can evoke interconnectedness instead of control and mastery: the drone can become a medium for planetarity.

During recent decades, the planetary has become an important concept in the humanities, including in media studies, cultural studies, and literature.[3] While globalism is rooted in the idea that the world is connected by the global flow of capitalism, the discourse of the planetary positions the human in relation to the planet as a zone of life and climate. The planetary view does not categorize the earth according to nation-states, international cooperation, or global networks, it sees the people on the planet as one, without making them the same. According to Gayatri Chakravorty Spivak, the term *planetary* stands in opposition to concepts of the worldly, the global, and the continental, which function according to the ideology of capitalism and remain imbued with cultural and national essentialisms.[4] By contrast, for *planetary subjects*, alterity is not derivative. In the positioning of human subjects vis-à-vis the planet, the conditions of differentiation are erased: "The planet is the species of alterity, belonging to another system; and yet we inhabit it on loan."[5] Planetary subjects do not insist on their property rights to the planet. They understand the earth as a place they have on loan and are thus positioned in opposition to the processes of global financialization and capitalist homogenization.[6] It is important to note that the concept of the planetary is also criticized for entailing total- izing and unifying tendencies that erase difference and heterogeneity, and my interpretations of the artworks strive to show this as well.[7]

This chapter argues that the drone's flattening gaze has the potential to envision planetary communities. In these visions of community, humans appropriate the earth not as a specific territory that belongs to some of them, but as a common planetary space that needs to be protected. In order to develop this argument, my first step is to historicize the drone, finding "family resemblances" between the drone and hot-air balloons. This media- archaeological and genealogical approach turns its face against techno- optimistic narratives of the drone as the latest game-changing technology and shows that the drone as the balloon's descendant has potential for creative, imaginary, and utopian visions of planetarity. Before I engage in a close reading of aesthetic drone imaginaries and their planetary visions, this filiation between the drone and the balloon must be mapped out.

Aerial Gas Balloons as Early Drones

At first glance, hot-air or gas balloons and drones seem very different. Bal- loons are beautiful, colorful airships that ascend serenely; drones are rather ugly. Drones have been developed and utilized mostly in the twenty-first century; gas balloons were invented in the eighteenth century. Neverthe- less, there are some striking similarities between the two aerial technologies. I do not see the invention of the drone as part of a linear, chronological narrative. Rather, I consider the aerial balloon as a distant relative to the drone, connected via a specific aesthetic mode of sensing and interpreting the earth.[8] Let us look at the beginnings of ballooning.

On June 5, 1783, only a few years before the French Revolution, the Mont- golfier brothers released their first hot-air balloon into the sky. From that day on, engineers, enthusiasts, entrepreneurs, artists, and writers experimented with, improved, developed, and wrote about balloon-flying technology. Félix Nadar, for example, was the superstar balloonist of the nineteenth century. His spectacular journeys and failures (such as the crash of his balloon *Le Géant*), as well as his early attempts to take aerial photographs from his balloons, still fascinate today.[9] Aside from stunning shows, aerial balloons also inspired fashion trends, influencing the design of clothing, fine china, and porcelain figurines.[10] Many late eighteenth- and nineteenth-century literary authors were enthused by balloons and wrote about these new airships with great passion.[11] Similarly to the emergence of contemporary drone art, balloon art became an aesthetic genre. Eighteenth- and nineteenth-century "balloonomania" in art and popular culture can be compared to the "drone-o-rama" of today's media hype around drone technology.[12]

As well as featuring in aesthetic works and popular entertainments, balloons, like drones, were seen from the outset as a dual technology that had strategic potential for warfare. Joseph Montgolfier himself noted, "By making the balloon's bag big enough, it will be possible to introduce an entire army, which, borne by the wind, will enter right over the heads of the English."[13] Like drones, balloons were immediately instrumentalized by the military for aerial reconnaissance, bombing, and the transportation of goods. For example, at the Battle of Fleurus on June 26, 1794, the French army used the balloon *L'Entreprenant* to get a better view of the enemy. Although the French won this battle—perhaps partly thanks to the balloon's "shock and awe" effect—France's military use of balloons was interrupted in 1799, when Napoleon disbanded the Aerostatic Corps. Nevertheless, French war balloons were airborne again during the Franco-Prussian War of 1870–1871. Balloons were also used as weapons during the American Civil War.

My attempt to demonstrate a family resemblance between balloons and drones might face the objection that a balloon is not a remote sensing technology. Balloons are crewed airships, while drones are crewless—indeed, an alternative name for drones is *unmanned aerial vehicles*. As I indicate in my discussion of the drone as cyborg in chapter 1, this concept of "unmannedness" is complex, because humans are closely enmeshed with drones and cannot be separated from the drone assemblage. A glance into the history of ballooning supports this idea by revealing a similar entwinement between humans and machines. The first balloons went into the skies entirely crewless. Later, to test the viability of human survival at atmospheric heights, animals (sheep and chickens) were sent airborne. It was only after animals had returned unharmed that humans took to the air. Many of these early crewed balloons did not move completely freely but were tethered to the ground by a rope. This development from crewless to "animaled," tethered, untethered, and eventually crewed flight undermines the very notion of "unmannedness." In the crewed balloon, a human pilot sits in the gondola and steers the vehicle by pulling ropes and controlling the ballast. In the drone, the pilot sits on the ground and controls the drone via mouse clicks. "Unmannedness" is thus a matter of relative distance and tethering, rather than of physical presence. The most important similarity between the balloon and the drone, however, is their flattening aerial perspective, which I discuss in the next section. Both drones and balloons are technologies of sensing that flatten the world. To demonstrate this, I take one of the very first accounts of ballooning as the centerpiece of my analysis because it provides precious insights into the aesthetics of flattening and its affectivities.

Flattening the World with Aerial Gas Balloons: Thomas Baldwin's *Airopaidia* (1786)

Thomas Baldwin's book *Airopaidia* (1786) gives a wonderful account of the early days of ballooning.[14] It contains notes recorded during his one-day aerial journey over Chester in the UK, with detailed technical information, such as atmospheric data, the weight of the balloon, and its materials. In one way, the text reads like a manual for aspiring balloonists, offering knowledge and instruction. But *Airopaidia* is much more than that: it is also a philosophical and poetic essay on the aesthetics and affectivities of ballooning as a new technology of seeing. The flattened aerial view from the balloon is a vital part of this. Baldwin writes the following about the unprecedented new perspective: "The endless variety of objects, minute, distinct and separate, tho' apparently on the same plain or level, at once striking the eye without a change of its position, astonished and enchanted."[15] The key words in this sentence are *plain* and *level*, which describe the balloon view as a flattening gaze that looks down onto the earth rather than toward the horizon. Formerly prominent features of the earth, such as hills, cliffs, forests, and villages, form an abstract pattern that is devoid of spatial depth; objects appear isolated and disjunct. This effect is also present in the remarkable illustrations in Baldwin's book.

"A Balloon Prospect from above the Clouds" is a flattened image of the earth from above. We see hills, valleys, waterways, and hints of the city of Chester. Clouds cover great parts of the earth's surface; hovering above them, the balloon cannot see through them. The aerial view does not convey any depth of focus, and the earth is flattened onto a colorful rectangular surface.

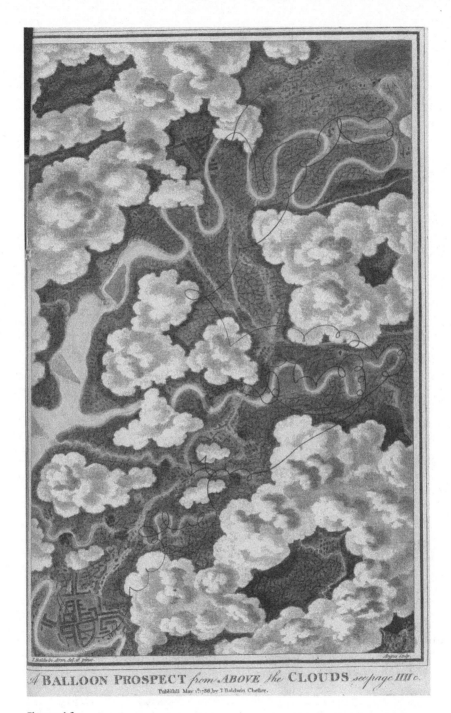

Figure 4.1
William Angus after Thomas Baldwin, "A Balloon Prospect from Above the Clouds,"
from *Airopaidia, Containing the Narrative of a Balloon Excursion from Chester, the Eighth
of September,* 1785, Chester: J. Fletcher, 1786, hand-colored etching. © Yale Center for
British Art, Paul Mellon Collection.

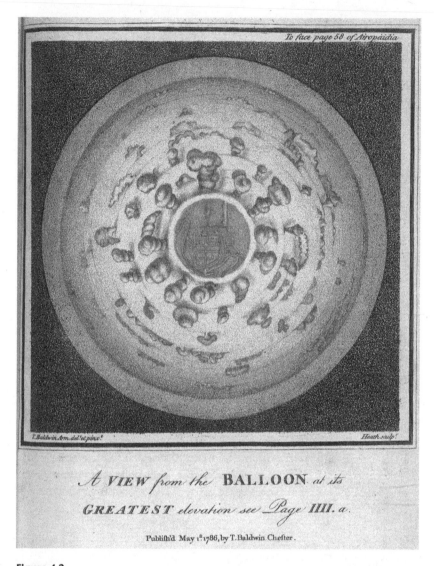

Figure 4.2
William Angus after Thomas Baldwin, "A View from the Balloon at Its Greatest Elevation," from *1785*, Chester: J. Fletcher, 1786, hand-colored etching. © Yale Center for British Art, Paul Mellon Collection.

In "A View from the Balloon at Its Greatest Elevation" too, the earth is a flat plate. Although the image suggests the globe of the earth, the top-down perspective flattens it, recalling a geocentric Ptolemaic view of the planets.

How can we interpret Baldwin's images of the earth? Why are they interesting in regard to the sensing mode of the drone? For me, Baldwin's illustrations are prototypes of drone images: they not only flatten the world but also contain an element of nonhuman vision, as products of a remote sensing technology. Baldwin's drawings suggest a mode of "operative visuality."[16] This type of visuality is found in "images for usage" (*Gebrauchsbilder*), such as graphs, diagrams, and maps.[17] Sybille Krämer roots this operative visuality in the quality of flatness.[18] It is important to note that operative visuality is not functionally equivalent to operative images in the military context.[19] As I outline in chapter 1, operative images are often discussed in relation to drone warfare, where the drone's information system is involved in making life-or-death decisions based on machine-readable data. The flattening effect of operative images has therefore been connected to power, aggression, and domination.[20] But that is not the case here. Baldwin's images are related to these operative images: they are aerial, have the quality of flattening, and obtain operative visuality. But they utilize this operative aesthetics to articulate interconnectedness, not to execute violence. They provide an alternative vantage point to understand the operativity of the aerial image in a more creative, aesthetic, and sensuous fashion. As Laura Kurgan states, it is important not to reduce the operative flattened images taken by satellites to the contexts of military violence; we must also engage with the aesthetic aspects of such images.[21]

It is this aesthetic side of aerial images with operative visuality that I want to trace in Baldwin's balloon images, and in the aesthetic drone imaginaries of contemporary artists. Although Baldwin's flattened images have cartographic features because they show topographical and geographical elements, they do not quite qualify as maps. The clouds suggest an atmosphere of serenity and peacefulness; the lines (marking waterways and valleys) have an ornamental and arabesque beauty; the many shades of green used to color the forests recall landscape images. Svetlana Alpers has questioned the distinction between maps and landscape images by analyzing seventeenth-century Dutch paintings that combine cartographic elements with landscape painting aesthetics.[22] Baldwin's images merge the cartographic with the aesthetic; drone images can similarly be characterized as ornamental, and as recalling nonfigurative art.[23]

Aesthetic beauty is emphasized in Baldwin's enthusiastic textual descriptions of the view. He describes his ascent in the balloon as "fantastic" and a "spectacle," and once he was in the sky he "shouted for joy."[24] Note the following passage about his tearful astonishment at the new perspective: "A tear of pure delight flashed in his eye! Of pure and exquisite delight and rapture; to look down on the unexpected change already wrought in the works of art and nature, contracted to a span by the new perspective, diminished beyond the bounds of credibility."[25] Being in the air is an experience of the "true sublime,"[26] an aesthetic category to which Baldwin frequently refers as he describes his journey. In doing so he takes the psychological approach to the sublime found in Edmund Burke's *Philosophical Enquiry into the Origin of Our Ideas of the Sublime and Beautiful* (1757).[27] Like Burke, Baldwin attributes the quality of the sublime to affects, feelings, and bodily sensations by mentioning that the aerial view from the balloon "goes beyond the bounds of credibility" and is beyond the "power of language."[28] Baldwin emphasizes several times that the aerial perspective triggers a bodily "dazzling delight"[29] and is an "overwhelming experience."[30]

How should we understand these flattened, sensuous images of operative visuality, as well as Baldwin's sentimental descriptions? We could interpret this view from the balloon as a superhuman, godlike view from above, a case of omnivoyance (see chapter 1). Baldwin's illustrations, however, do not suggest that he could see everything. The images show only flat fragments of the earth, and clouds cover parts of the land—not transparent Olympian clouds, but obstacles to the view. The view from the balloon is not entirely human either. There is no identifiable viewer, as there would be with a central perspective or first-person view. Instead, the flattened images suggest that it is the balloon that does the seeing. The partialized view seems neither godly nor human and is more like the view of a nonhuman agent (perhaps similar to Julius Gustav Neubronner's pigeons, discussed in chapter 1). The flattening perspective from the balloon grants Baldwin a nonhuman perception and experience of the world. The image is thus the product of the balloon itself as a technology of seeing. Indeed, it is reminiscent of Odilon Redon's lithograph *The Eye Like a Strange Balloon Mounts Toward Infinity* (1882), which portrays a balloon as a gigantic eye looking up into the sky. In this image, the eye is dissociated from physiological

Figure 4.3
Odilon Redon, *The Eye Like a Strange Balloon Mounts Toward Infinity* (1882). © The Museum of Modern Art, New York Scala, Florence

sight and transferred to a nonhuman context by means of optical technology. Although Redon's eye-balloon looks up while Baldwin's gaze is directed downward, in both cases the balloon's construction is clearly made into a technology of seeing.

The balloon's nonhuman gaze is highly dependent on the forces of nature, specifically on the atmospheric conditions, climate, and meteorological situation. Note the fine black curly line in Baldwin's image "A Balloon Prospect from above the Clouds," which indicates the balloon's whimsical route. This route (and the possibilities for perception that it provides) is not entirely controlled and steered by a human pilot; winds, temperature, and air pressure all play a part. Thus the balloon as a technology of seeing does not have the human at its control center, but instead is steered by nature. In this connection Baldwin makes the striking remark that the "scenes" the balloon delivers are "heightened by the pencil of nature."[31] This is interesting from a media-historical perspective, as the notion of the "pencil of nature" is best known from William Henry Fox Talbot's eponymous illustrated photobook, published in 1844–1846. Talbot described photography as the art of photogenic drawing, in which it was mostly nature (not the artist) that was the creator of beauty. Joanna Zylinska sees Talbot as an early demonstrator of the nonhuman in photography, since he explored photography as a technology that was predominantly determined by rays of sunlight.[32]

Baldwin's work defines this nonhuman gaze through its interconnectedness with the earth. The balloon view triggers moments of ecstasy when the pilot no longer measures, controls, or masters the balloon but is taken over by an overwhelming feeling of joy and delight. In these moments, the subject and the world are no longer separated, and the boundaries between them become fluid, permeable, and contingent. When Baldwin is airborne, he speaks of the light that connects him to the earth and the heavens: "Rays of light darted on the eye as it glanced along the ground: which, tho' of a gay green color, appeared like an inverted firmament glittering with stars of the first magnitude."[33] His narrative voice fuses the earth with the cosmos in a Romantic vision of spiritual union. This noninstrumental merging of human and nature aligns these flattened aerial images with the idea of planetarity that I detect in the drone imaginaries of contemporary artists. In Baldwin's aerial experiences, the seeing subject forms a nonbinary constellation with the planet. One could characterize Baldwin's planetary imaginary in terms of Bruno Latour's trope of Gaia, a vision of the earth as a "superorganism"[34] of beings and materials that cannot live apart, and from which humans cannot extract themselves. Gaia goes beyond the bifurcation of nature and culture and is defined by a connectivity of the human and the nonhuman; it thus articulates a critique of the notion of earth as a territory that needs to be measured, controlled, and dominated.[35] The planetary dimension that is evident in Baldwin's discourse on aerial ballooning articulates an affective, noninstrumental relationality between this technology of seeing, the environment of the earth, and the human subject—a relationality that engenders this chapter's trajectory of discussion regarding the planetary communities of drone imaginaries. In contrast to my previous chapters, which at times define community on the basis of an individual human–machine interaction, this chapter extends the idea of community to a planetary scale. This means that I explore imaginaries of communities that entail a meta-individual level and contain a vision of how humans live together on the planet and interact with it via technology. As I will show, these planetary communities can exceed the idea of national boundaries and territories. In the remainder of this chapter, I show that the civilian drone has similar planetary capacities enabled by its flattened sensing of the earth, and that this is different than the flattening operation of the military drone. To specify the difference between these two traditions of flattening, I first offer a brief examination of the military drone from an artistic point of view. Therefore, my next example serves as a counterfoil to the duality of flattening.

The Flattened View of the Drone

Trevor Paglen is well known for his photographs and films about mass surveillance, data mining, and military security technology. For example, his collection and exhibition of official and semiofficial military patches worn by US soldiers revealed secret visual codes and clusters of signification within military discourse.[36] He also engaged with the topic of military drone operations by making several photographs named *Untitled (Reaper Drones)* in 2010. These images show colorful expanses of open sky, in which we can barely make out the almost invisible traces of a Reaper drone. Paglen's video *Drone Vision* (2010) is particularly important for my argument, since it exhibits the flattened vision of the military drone, indicating a nonhuman form of vision. The video exploited a security flaw in the transfer of images from drones to a US-based pilot via unencrypted satellite uplinks. The source material for the video was intercepted by an amateur hacker from an open channel to a commercial communication satellite over the Western Hemisphere. The five-minute film mostly shows aerial images from the perspective of the drone, which hovers in the sky filming mountains, roads, trucks, houses, and clouds.

Paglen's film begins with aerial shots taken by a drone. It shows the drone circling one specific point and then zooming in to the surface of the earth, revealing signs of mountains, roads, and infrastructure. These satellite images

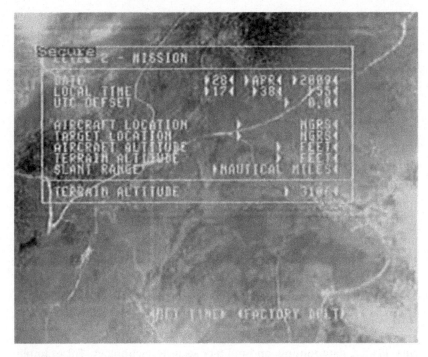

Figure 4.4
Video still from Trevor Paglen, *Drone Vision* (2010). Archival pigment prints. 16×20 in.
© Trevor Paglen. Courtesy of the Artist, Metro Pictures, New York and Altman Siegel,
San Francisco.

exhibit a flatness: they show the earth in a maplike fashion, with no central perspective or vanishing point. They do not entail a vision of planetary
community, but rather construct a community of potential enemies. Very
differently than Baldwin's aerial view, the flattened gaze of the drone is connected to operations of violence. Visible at the images' margins are technical
metadata about the time, location, aircraft, altitude, and mission, which can
be sorted through by data analysts and algorithms. Possible individuals in
the image are turned into dots, patterns, and clusters as signification points
for targeting. Thus *Drone Vision* shows operative images that establish a kill
grid, decide about life and death, and dehumanize the subject.[37] According
to media theorist Bernd Siegert, who has discussed the grid as a cultural
technique, the flattening is key.[38] The grid "is an imaging technology that,
by means of a given algorithm, enables us to project a three-dimensional

world onto a two-dimensional plane."[39] Thus the grid represents a strategy of abstraction that controls individuals via quantitative systems. What is in focus in the grid is not individual subjects, but a datafied public constructed by algorithms and data analysts.[40] Although the flattened drone vision seems to undermine the hierarchical and verticalized scopic of the military gaze, it nevertheless suggests a gaze of sovereignty. The gridlike structure exerts a power relation between those that construct the grid (the drone and its human agents) and those that are captured by this imaging process. Thus the drone suggests a vision of a community of potential enemies from which the drone's pilot and the image's viewers must be protected. Paglen's *Drone Vision* performs an aestheticization of these gridlike operative images, and in doing so makes them into an object of study and critique. The film distorts the power of operative images by experimenting with disruptions (static), inserting the image of an analogue clock, and portraying the drone as an animated machine with an endless appetite for data.[41]

But here I wish to trace another dimension of the drone's aerial flattening, beyond the grid, by focusing on artists whose work engages with amateur drones. As established earlier, I approach the flattened view of the drone from the Romantic perspective of ballooning, which provides an alternative vantage point to see and sense the earth. I therefore also choose drone artworks that directly or indirectly trace the relationship between the drone and the balloon, and which demonstrate that flattening can have a planetary dimension. Indeed, it would have been interesting for me to include Paglen's *Last Picture Project* (2012), for which he crafted a disk microetched with a hundred photographs of the earth and its civilizations and beamed it into space via satellite. However, an important criterion in my selection of artworks was that the drone had to be the central medium of the aesthetic process. Moreover, I am interested in artists that are somewhat less canonical in the discourse on surveillance art, and whose works engage with the amateur drone, offering new interpretations of its aesthetic sensing capacities. Therefore, I have chosen visual artworks by Tomas van Houtryve, the artist collective Postcommodity, and Ignacio Acosta. None of these artists treat the drone's flattening gaze as calculated, operational, and datafied; instead, they focus on its ability to envision planetary communities. I am therefore going to suggest that drone artworks by Houtryve, Postcommodity, and Acosta can be seen as manifestations of planetary art.

In his analysis of planetary art, Terry Smith defines "world scale regimes of seeing" as one aesthetic configuration of the planetary.[42] These regimes of seeing are particularly enabled by "new technologies of vision and visualization"[43] such as drones, social media, and digital imaging. World-scale regimes are not confined within national or multinational frames; they encompass the whole world and planet. Thus my approach also relates to Paglen's statement that drones create their own "relative geographies, folding several noncontiguous spaces around the globe into a single, distributed, battlefield."[44] It is interesting that this "relative geography" can be understood beyond the military context, as a single, distributed, *planetary* perspective. In other words, we are dealing here with similar visual features and visualized registers, but utilized in another field.

My readings of the artworks will show that the drone can invoke visions of community based on this planetary perspective. These communities are not focused on specific territories, or on global and local perspectives, because the planetary configuration of seeing embraces all: the universe, the planet, the flora and fauna of the world. This planetary vision of community is at stake, as I will show, in individual aesthetic drone imaginaries that engage with aerial perceptions of the earth, as well as in the drone's function as a tool of protest in contexts of political activism and intervention.

However, these planetary communities should not be seen as efforts to homogenize all of life into a one-world totality (in this sense I would say they differ from Gaia). Rather, the planetary must be understood as shifting, paradoxical, incomplete, constantly open to reconfiguration, and prone to integrate concepts of alterity.[45] It is important to keep the shifting nature of the planetary in mind as part of my readings of individual artworks. Although these aesthetic imaginaries can evoke utopian perspectives on planetary communities, they also reflect the precariousness of such visions. And it is precisely the drone as surveillance technology that can undermine and subvert those visions. These drone artworks evoke self-reflexivity about the military affiliations of the amateur drone and make it into a medium of political critique.[46] In other words, these examples of drone art demand a multiperspectival mode of interpretation where meaning oscillates without becoming fixed. In these drone artworks, the planetary dimension can emerge, but, as I will now show, it is also bound to a dialectic of appearance and disappearance.

Tomas van Houtryve's *Divided* (2018)

Houtryve is a conceptual artist, photographer, and author. He creates "using a wide range of processes, ranging from 19th century wet plate collodion to thermal imaging and Augmented Reality. His projects challenge our notions of identity, memory and power, often by highlighting the slippage of war-time structures into everyday life."[47] He has also engaged with the drone as a photographic medium. For example, Houtryve's photographic series *Blue Sky Days* (2014) shows images shot by a drone in domestic settings, revers-ing the military perspective of the drone and directing it at US territory.[48] My discussion here focuses on his video installation *Divided* (2018), which connects the flattened drone view to representations of the earth.

Divided was shot by an amateur drone over Baja, California, where Mex-ico and US territory meet and a steel fence extends the border into the Pacific Ocean. The film suggests a nonhuman element of vision: not only is it a product of a remote sensing technology, but it also shows spaces devoid of humans. Recalling the balloon view, the film begins with a fixed bird's-eye view onto the waves as they crash perpendicularly into the barrier. The drone does not move; it hovers above the ocean. We see a static image with moving waves; birds fly through it, reminiscent of the drone that hovers above them. In ways similar to Baldwin's illustrations, the image suggests a flat topography of the earth. There is no central perspective, and the film seems to display an abstract surface. We are no longer sure whether waves are waves; the white and gray pattern could equally be an image of a moun-tain formation, shot from a point high in the sky. Or it could be a type of microvision, an enhanced close-up of the surface of an organic structure. Like images made by microscopes, the drone footage highlights texture and materiality. Houtryve's installation plays with scale and makes the surface of the distant ground look close and magnified. The sensorium of the drone is also more than visual: *Divided* presents the sounds of the ocean waves. One hears the roaring and crashing of the surf, the rustling of the wind, and the rumbling of the water. The ocean in Houtryve's film, and the repetitive sound of the waves, imposes a different sense of time, suggesting eternity. Time is not an instrumental measure, but supports the idea of continuity.[49]

In the course of the film, the static image and soundscape of the ocean's surface eventually becomes a moving image, and the landscape glides beneath the remote aerial drone camera. After a while, the top-down drone

Figure 4.5
Still from Tomas van Houtryve, *Divided* (2018). © Tomas van Houtryve.

camera tilts to capture the shore and the mainland. We see US territory
with roads and meadows, and the more populous Mexican side with its
sports stadium, houses, roads, cars, and trees. Although at the end of the film
the perspective runs toward the horizon, there is still no central focal point.
Even when the drone view reaches the horizon line, the latter appears like an
artificial drape as the film whitens and fades out. Houtryve has noted that
he comes from still photography but became intrigued by moving pictures
thanks to the new visual technologies available on cellphones and drones:
"Still photography is about seizing a moment in time. But now you can see
a little bit before or a little bit after that moment, or maybe the photo can
even magically come to life. There is an overlap between still and moving
images."[50] Reminiscent of the view from a gliding aerial balloon, the moving
photograph gains a visual intensity: it suggests an atmosphere of fluidity. This
fluidity is crucial to the imaginary of community that is at stake in *Divided*.

What visions of community does *Divided* suggest? The drone's flattening
gaze conveys an impression of the land's wholeness. Although the waves are
divided by a border, they are shown to be the same on either side. Houtryve's
film suggests the interconnectedness of the earth. The military drone is
often seen as superseding borders, waging war in areas where it is not held
to account. But here, the drone view reveals the continuity of landscape,
surface, and earth, fusing together territories that have been artificially

separated. Like Baldwin's balloon, the drone gaze in *Divided* embodies a planetary perspective, suggesting that the world is one, connected by water, and belongs to everyone. In Houtryve's work, the drone hovering over the ocean exudes this planetary perspective as it highlights the artificiality of political boundaries by flattening them. *Divided* thus suggests a vision of a human community that is not separated into different groups or nations, instead seeing the earth as a planetary space for all. Notably, the drone focuses on the unifying element of water. In planetary studies, oceans and water are often used as a metaphor for the oneness of the world.[51] Water evokes a Neptunian worldview: water is life affirming; we all came from the water onto the land. Water is fluid, constantly changing, transient. It seems to carry a fleeting planetary energy, a power to transcend national and territorial boundaries. It embodies an undivided natural space, and the life that came out of the water is not divided or demarcated by oppositions; origins no longer apply. This water-alterity is not derivative, and ancient aquatic life-forms as planetary creatures embody an inexhaustible form of difference. Thus the planetary offers a "para-rational image"[52] of the earth, suggesting the earth as a community for all. One might be tempted to connect this community to a primitive nostalgia or esoteric totality. But *Divided* does not suggest that we are all one and the same. The fluidity of water offers a metaphor for thinking difference beyond the "other," thinking communities in nonessentialist ways. This echoes Jean-Luc Nancy's ideas about nonidentity with respect to community: Houtryve's *Divided* reflects upon the question of whether communities should be understood solely in terms of national identities and territories, or whether instead the humans on this planet are all connected by common land and water. Like Houtryve's photographic series *Lines and Lineages* (2017), *Divided* reflects on the arbitrariness of borders, homelands, and national identities, and as its title suggests, it questions national divisions.

However, Houtryve's utopian imaginary of a planetary community in *Divided* is transitory and dialectical, since it can be undermined by the drone itself. By following the border, the drone's camera eye organizes the earth into a grid that masters, controls, and rules the once-unified space into separate territories. Similarly to Paglen's flattened operative image with its metadata, Houtryve's film shows how the drone as a technology of seeing can lose its planetary dimension by following a boundary line. The drone can become complicit with the grid, dividing people into separate territories. It recalls

the "atmospheric policing" of the Mexico/US border, which is heavily surveilled by air using drones, satellites, and airplanes.[53] *Divided* thus reveals
the drone's dialectical power in its envisioning of human communities: its
utopian dimension of planetarity, and its dystopian dimension of geometriz-
ing, ordering, ruling, and dividing. Houtryve's work oscillates between these
dimensions, without arresting the process of interpretation on either side.
This hermeneutic complexity is also at stake in my next example.

Postcommodity's *The Repellent Fence* (2015)

The artist collective Postcommodity launched their aesthetic landscape
installation *The Repellent Fence* in 2015. It was a binational art project in the
US/Mexico border area, near Douglas, Arizona, and Agua Prieta, Sonora.[54]
The Repellent Fence comprised twenty-six tethered helium balloons across a
stretch of two miles, roughly one mile into US and one mile into Mexican
territory. The location of this temporary artwork was decisive, since the
place where it was erected encompassed the tribal lands of Apache, Tohono
O'odham, and other Native peoples. This contested landscape, where thousands of people cross the national border legally and illegally, is heavily
surveilled by the US Border Patrol.[55] Native communities on both sides of
the border have been affected by the realities of US politics—particularly
during recent years, with the Trump administration's initiative to build a
wall along the border between the US and Mexico. The aerial balloons of
The Repellent Fence hovered a hundred feet above the Sonoran Desert, creating a permeable line. The balloons were ten feet in diameter, and their
bright colors could be seen from afar because they were scare-eye balloons designed to keep insects and birds away from plants. Interestingly,
the balloons deployed medicine colors and iconography (bird's eyes) used
by Indigenous peoples from Latin America to Canada. This connection
between Indigenous iconographies of eyes in the sky and a repellent fence
clearly constituted a protest statement against US immigration politics.

The ephemeral character of this artwork (the atmospheric movements of
the balloons, and their shaping of a permeable border), as well as its place-
ment on contested territory, spoke to the interwovenness of the history of
Indigenous people, the US, and Mexico. The line did not function like a border but traversed nation-states, evoking tribal areas and colonial histories.

Figure 4.6
Through the Repellent Fence. Photo © Micheal Lundgreen, courtesy of Postcommodity and Bockley Gallery.

The Repellent Fence crucially reveals the nonscopic sensing of the amateur drone, not only because it engages with the aerial gaze of balloons but also because the Postcommodity collective used an amateur drone to film the balloons and their environment. As I will show, this nonscopic sensing also generates imaginaries and actual formations of communities. The project is documented in the film *Through the Repellent Fence*.[56] In addition to showing the cultural, political, and aesthetic contexts of the installation, the film includes plenty of drone footage of the surrounding landscape and the sky around the balloons. The Postcommodity website also displays drone images of the installation.[57] Following the line of the balloons, the drone films the vastness of the desert. We see the desert valleys, dry mountains and hills sprinkled with greenery, and dried-out riverbeds. The footage conveys a flattening effect as we see the earth's surface and geological formations via the drone. Again, these flattened images of the earth, produced by the drone as a nonhuman technology of sensing, provide a planetary view. Like the ocean in Houtryve's film, the divided land in *The Repellent Fence* appears as a single geography with shared fauna and flora, highlighting the randomness and artificiality of the border. The drone not only connects to the ground, it also encompasses the atmosphere: it gets close to the balloons, shows how they hover in the wind, and almost seems to dance with them. Indeed, the balloons in this artwork were staged as a ceremonial performance,[58] and the sound and music are important components of the film on the artists' website as well as in the documentary. Again, the drone sensorium displays its synesthetic features, combining visuals with sound. The sight of the balloons in the air, the music, and the drone images of the earth and sky convey an aesthetic sublimity that recalls Baldwin's ecstatic reactions to the aerial view. We perceive the balloons not as a scary fence, a wall that repels, but as festive, exuberant, and celebratory.

Roberto Bedoya, executive director of the Tucson Pima Arts Council, aptly described the planetary aspect of *The Repellent Fence*: "Land is not just property we own but space that demands stewardship as a placemaking, place keeping activity that acknowledges what is sacred in the land via vistas, ceremonies, song and care. This worldview is an obligation embedded in the sovereignty of context, a form of governance that the *land* asserts."[59] *Stewardship* is a key term in planetary studies, often understood as an ethical caring for the planet and environment. Humans should act as stewards, not exploiting, ruling, and managing the earth, but taking care of it in sustainable and ecological ways.[60]

This care and stewardship are closely interlinked with an idea of planetary community. The drones and balloons show that the land does not belong to one nation or one tribe. Postcommodity stated explicitly that they were working in a trans-Indigenous fashion, addressing the history of all tribes without being specific about particular tribes. The fence of balloons stitched together the people in the area, interconnecting them and showing that they belonged together by remaining different. Thus the vision of communities engendered by the drone and balloons in this installation suggests Rosi Braidotti's vision of the social bonding that can be described as "We-Are-(All)-In-This-Together-But-We-Are-Not-All-One-And-The-Same."[61] Likewise, in the planetary community of Postcommodity, no one is excluded by national, tribal, or international boundaries: all humans belong to the earth and the planet. Again, as for Houtryve, this community vision does not enforce a totalizing homogeneous whole, but rather, as in Nancy's philosophy, is thought from difference, singularity, and diversity.

Besides outlining this utopian vision of a planetary community, *The Repellent Fence* also offers a concrete dimension of local community and neighborhood building. The founders of Postcommodity strongly engaged with local groups and institutions around their art project.[62] They spent years working with regional and national authorities, such as the US Border Patrol, the US and Mexican embassies, and the respective municipalities, in order to get the project underway. In addition, the performance integrated many local residents who helped with logistics, setting up the balloons and coordinating the project. These local communities included members of different tribes and nationalities. This shows that the drone can play an important role in uniting and organizing political activist groups—an observation that recurs throughout this book.[63]

However, the planetary communities envisioned by *The Repellent Fence* and its drone are ephemeral and temporary. Although drones and balloons alike embrace a nonscopic planetary sensing, the violent associations of the aerial gaze are still visible. In *The Repellent Fence*, the balloons lose their festive context and become complicit with rationalization, separation, and isolation. As technologies related to military reconnaissance, balloons and drones embody the constant aerial surveillance of this land, and the birds' eyes on the balloons symbolize the eyes of the predator—the US Border Patrol. The openness and permeability of the balloon line also refers to the invisibility of digital fencing and border surveillance in the area. Not just the ground but the whole of the atmosphere and sky are subjected to militarization and colonialism.[64] Thus *The Repellent Fence*, like Houtryve's *Divided*, engages in a dialectical process of critique: the aerial ensemble of balloons and drones gives us a glimpse of planetary community, but at the same time it also reflects on its own complicity as a means of surveillance. Nevertheless—and this is typical of drone art—the artists manage (at least for a moment) to recontextualize the drone for their own purpose and mission, making it into a tool of critique and political intervention. Such recontextualization also occurs in this chapter's final example of drone art.

Ignacio Acosta's *Drones and Drums* (2018)

The work of contemporary artist and writer Acosta engages with the ecological effects of mining and land exploitation. His book *Copper Geographies* depicts how copper has shaped the power structures of the modern world.[65] By virtue of its mining and usage in military weapons and cellphones, copper has an impact on our environment and the societies in which we live. Acosta's multimedia art projects also revolve around the exploitation of nature and its effect on human communities. His video/sound installation *Litte ja Goabddá* (*Drones and Drums*) (2018) artistically documents a local community of resistance against big-industry profiteering. The film is about the Indigenous Sami people around Kallak in the Swedish county of Norrbotten, and their fight against mining companies' exploitation of the land. Amateur drones play a decisive role in the film: Acosta uses them as visual and sonic media, and the Sami themselves deploy commercial drones as tools of protest. Again, the drone sensorium opens up to more-than-optical modes of sensing, integrating sound, rhythm, and music.

Drones and Drums consists of two channels, with screens showing two different films at the same time, enabling simultaneity effects in its visual and sonic presentations. The film is structured into four chapters, entitled "Forest," "Water," "Iron," and "Resistance." The title is displayed at the beginning of each chapter, first in the Sami language and then in English translation. The film begins with a prelude that shows a man standing in a vast, snowy forest landscape in the Sami region. The man wears traditional Sami dress and plays a drum. Drums are an important part of Sami culture, since shamans (*noaidi*) use them for spiritual purposes, such as attaining trance states and communicating with ancestors.[66] The drums are often ornamented with symbols and other features portraying the location of the family in question, reindeer hunting, and cosmological signs. In this opening sequence, the drumming Sami man is joined by a drone controlled by another man wearing traditional Sami dress, and the changing channels show the drummer, the pilot, and the drone at the same time. The sounds of the drum and the sounds of the drone overlap, creating an ancient and hypermodern soundscape that hovers above the land. The drone and drum evoke a prediscursive, mythical connection to the earth, and this is further enhanced by the film's flattened drone images.

Acosta's flattened aerial images embedded in musical soundscapes are of the greatest interest for my purposes. As with *Divided* and *The Repellent Fence*, this remote and nonhuman form of vision demonstrates a specific connectedness to the earth.[67] As the drumming and drone sounds intensify during the first minutes of the film, the drone camera shows flat images of reindeer

Figure 4.7
Still from Ignacio Acosta's *Litte ja Goabddá (Drones and Drums)* (2018). © Ignacio Acosta.

Figure 4.8
Drum from Lule Sami area, with hunting motifs (ca. 1673). Public domain.

farming shot from above. The pattern of the reindeer herd in the flattened drone images coincides with the abstract designs and cosmological semiotics of Sami drums. In this scene, the flatness of the drone images is not connected to the violent operative gaze of military drones. Combined here with specific soundscapes and the patterns of the herd, the flatness instead embodies an interconnected sensing of the earth. And in this case too, the sensorium of the drone evokes a planetary vision of community. The

Figure 4.9
Still from Ignacio Acosta's *Litte ja Goabddá (Drones and Drums)* (2018). © Ignacio Acosta.

flattened images of reindeer are associated with ancient rites of hunting and gathering, when hunting entailed not maximizing profit but maintaining a subsistence economy. By drawing a parallel between the abstract patterns on the drums and the abstract patterns of the herd, the images echo an ancient cosmological order of the earth, before its modern instrumentalization and exploitation. They introduce the dimension of the planetary, not as a call to premodernity but rather as enmeshed with the drone's high-tech sensing modes. The flattened images form a critical intervention against profiteering on the land. This is especially key in the film's last chapter, "Resistance."

"Resistance" contains many images made by drone that document the effects of clear-cutting in the forest. The drone films the disappearance of the forest, the fading grass, the erosion, and the residual stumps of trees. Thus the drone functions as what media scholars have called an *eco-medium*, both registering environmental damage and tracing the materiality of media components.[68] The drone then becomes a means to form a community of activists: the Sami group use it as an eco-medium to document their protest. "Resistance" makes this explicit by presenting drone footage shot by Sami protesters against logging and mining. We see Sami people gathering in the forest and building barricades against the bulldozers. The Sami group steer and control these drones; the drones make the Sami voices heard. Thus the Sami have reclaimed the drone, which works for them as a tool of protest.[69] In this respect, Acosta's work is also close to the video installation *Eyes in the Sky* (2017) by Frédérick a. Belzile, which is further discussed in chapter 6. Belzile's work records the use of drones by Sioux protesters at Standing Rock, North Dakota, in 2016. These protesters were opposing the construction of a pipeline that put their land and cultural life at risk. The drone footage became an important tool to document the scope of the protest activities, as well as the authorities' use of force against the protesters.[70]

Acosta's film too articulates a protest community by means of the drone. But it is important to grasp that the message of his film and research project is not simply to reinforce the rights of this specific group of Sami. Rather, and again, similarly to *Divided* and *The Repellent Fence*, the drone and its flattening gaze upon the earth envisions a planetary community that is not defined by the oppositional binary logic of Indigenous versus non-Indigenous, nature versus technology, ancient versus modern. The sensorium of the drone, its soundscapes and aerial images, collapses these dichotomies and suggests that the earth should not be owned by anyone. In this vision of a planetary community, the human is not the sole controller of the land, but lives in a sustainable relationship with the natural environment.

Yet the drone does exhibit dialectical complicity with the exploiters. The Sami land is also the location of a large military base, and Acosta's drum-drone sensorium evokes associations with military surveillance. As a technology of military origin, the drone reverberates with the history of the colonial oppression of Sami people, and of Swedish attempts to mission-ize the Sami and convert them to Christianity. However, the drone-drum soundscape and its flattened images do not foreground this violent aspect of Sami history. The drone is portrayed not as a penetrating intruder but as an aesthetic, spiritual, and political medium. Acosta's drone does not maintain its alliance with the scopic gaze of military imperialism; its aerial view opens a perspective on the vulnerability of the earth, showing the wounds inflicted by clear-cutting and its impacts on the local Sami community. As Sarah Tuck states in her analysis of *Drones and Drums*, the drone "can be understood as a critical visual method of thinking and seeing in alliance, one that challenges the aerial view as a colonial perspective."[71] The imperial gaze of aerial vision resonates in the film, but it is then reappropriated for a planetary context. It is precisely this creative recycling of drone technology that conjoins *Drones and Drums* with the other drone imaginaries I have discussed in this chapter. By hijacking the drone for their own aesthetic purposes, all of these artworks suggest utopian prospects for nonexclusionary planetary communities.

Planetary Communities

The artworks analyzed in this chapter, which all engage with the amateur drone as an artistic medium, shed light on flattening as a way in which drones sense the world. Drone flattening has mostly been discussed to date in military contexts, where it is seen as a feature of the operative images that facilitate kill chain operations in drone strikes. But in this chapter I have shown that this is not the whole story. The artistic experiments by Houtryve, Postcommodity, and Acosta have disclosed another dimension of flattening, one that articulates a state of interconnectedness with the earth. My historicization of the drone in relation to the aerial balloon has helped me to unearth this other dimension of the aerial perspective.

Nonhuman modes of remote sensing and image-taking, which are key to both aerial balloons and drones, enable this planetary dimension. Instead of perpetrating the violence of the grid, the sensuous registers of aesthetic drone imaginaries discussed in this chapter generate visions of planetary community. In these ephemeral and utopian visions, the human is no longer the master, dominator, and exploiter of the earth, but encounters the environment through care, stewardship, and sustainability.

In this respect, these artworks give us hints about what we should do in the world in light of the exploitation of the earth. For example, Acosta's work clearly conveys an activist political message against clear-cutting and profiteering from the land on which the Sami people live. But the artists also combine this ethical component with a political vision of community. Perhaps most significantly in *The Repellent Fence*, the drone suggests communities that are not built on national essentialism, exclusion, or territorial claims. This vision of community embraces the inhabitants of the whole planet and undoes the conditions of alterity altogether. This erasure of alterity does not mean that there are no longer any differences. There are still differences. However, they are understood not as distinctions through national identities but more, as Nancy suggested, as singularities. Nor are the utopian visions of planetary community that these drone artworks evoke confined to the realms of fiction and aesthetic imagination. Acosta and Postcommodity both root their artworks in communities of political activists, and they understand their art projects as political interventions. The drone—a technology commonly associated with violence and oppression—can not only display an alternative mode of sensing but can also be reappropriated as a medium of political protest.

5   Volumetric Sensing and Postcarbon Communities

"Environment is not the ground or fundamental condition against which sensor technologies form, but rather develops with and through sensor technologies as they take hold and concresce in these contexts."[1] This statement by Jennifer Gabrys suggests that our sensor-based monitoring of the earth is making new environments that intertwine technology, nature, and people. Through sensors, humans connect to the earth not only from outer space but also on and beneath its surface. These sensual technogeographies, which entwine the human and the nonhuman, are the focus of this chapter, which analyzes the drone's ways of sensing the earth and enmeshing us with it. As we saw in the previous chapter, drones can flatten the earth with their aerial view from above. But flattening is not the whole story: earth-sensing drones can also go three-dimensional. Tracing this three-dimensionality of the drone sensorium, this chapter investigates consumer drones that perceive the earth in the contexts of agriculture, landscape design, and art. I call these *earth drones*, since they are equipped with special sensors that monitor, observe, and surveil the earth. They are interfaces and data processors that can look up from the ground to the sky, monitor the surface (soil) and its in-between spaces, and sense deep into the ground. I characterize this multiperspectival sensorium of earth drones as a form of "volumetric sensing"[2] that grasps the three-dimensionality of space. As Ole B. Jensen notes, "Volumetric thinking means an awareness of an x, y, and z axis that spans out three-dimensional space. Spoken more plainly, the fact that the drone can move up, down, sideways and to the front and back is all about the volumetric space in which it can move."[3] Accordingly the volumetric does not embody a static vision of three-dimensionality but is based on mobility pertaining a constant change of perspective.

How do earth drones and their volumetric sensors define our relationship with the earth? Do the military origins of drone technology determine this relationship, or can earth drones generate a connection to the earth that goes beyond the power relations of the vertical? What do these drones tell us about the earth in the context of the Anthropocene? And how does the earth drone's sensorium generate imaginaries of communities? To find answers to these questions, I use the term *volumetric* in a slightly broader sense than is usual in the fields of mobility studies and human geography, although I have been greatly inspired by research in those fields.[4] The fields of architecture, landscape design, and urban planning have used the concept of *volumetric sensing* for some time. In these instances, *volumetric* points to the idea of building up into the sky and grasping the three-dimensionality of urban space. Lidar and point cloud modeling are often applied for these purposes. The term *lidar* was originally an acronym for *light detection and ranging*, a remote sensing method that uses lasers to measure spatial ranges. A point cloud is a set of data points that can represent the three-dimensional shape of an object. In the context of these sensing technologies, the key question for me is what happens to our perception of the earth when volumetric sensor technologies are attached to a drone—when a drone camera sees in a way that creates three-dimensional spaces rather than flattened views from above.

I trace the volumetric dimension of the drone primarily in aesthetic contexts, an angle that recent research has left unexplored. Contemporary drone artists are experimenting with the volumetric sensorium of the drone, and such artworks are my focus here. Besides these artistic negotiations, the sensorial quality of earth drones (their *aisthêsis*) is also important, and I focus on how drones register, compute, calculate, and surveil the earth in nonartistic contexts, such as agriculture. My main intention is to show that earth drones sense volumetrically, and that in doing so they can generate imaginaries of communities that are in close dialogue with the discourse of the Anthropocene.

The term *Anthropocene* signifies the time period during which humans have affected the earth.[5] Recent discussions have extended this notion beyond its use in earth sciences; the humanities have used it to initiate critical debates about the human subject's role and position with regard to the earth, its sustainability, and the exploitation of the natural environment.[6] For me, the Anthropocene describes the slow realization that the human species does not ultimately own or control the earth. The human era is finite; we are destroying our own living conditions and wreaking environmental damage. Thanks to global warming and our exploitation of resources, the era of the human as the sovereign of creation may be nearing its end. This potential ending brings with it a change of consciousness, perhaps best expressed in Timothy Morton's conception of *ecology without nature*.[7] For Morton, "nature" is a construct that fuels the differentiation between civilization and the natural world, a dichotomy that instrumentalizes nature as an object of exploitation. Instead, he argues for a type of ecology that embeds the human and the environment in less hierarchical entanglements and relations.

One of my key interests in this chapter is to establish what role the drone's volumetric sensorium can play in this respect, and in what way it can shape imaginaries of communities on and with the earth. To put it more simply, I want to find out what we can learn from the drone about the earth. The drone can observe, register, and archive the destructive force of humans' impact on the earth. In this way the drone can "soften" the dichotomy between nature and civilization and demonstrate our close entanglements with the environment. Together with artistic, scholarly, and public discourses, the drone can critically reflect on environmental damage and our exploitation of the earth, and can make us think about the dichotomy between nature and civilization, which Morton sees as symptomatic of the hierarchical power relations of the Anthropocene. Thus the earth drone constitutes a critical seismograph of the Anthropocene and a medium that articulates postnatural (utopian) visions of communities. But again, we must be aware of the dialectics of the drone. This Janus-faced technology also exploits the earth, and in doing so it enforces that very division between nature and civilization. I will begin with its dark side.

Dronoculture, Precision Farming, and Multispectral Imaging

During the last decade, agriculture has been shaped by smart technologies, such as remote-controlled tractors, weeding robots, tracking chips for animals, and data-collecting sensing technologies. According to engineers and businesses, smart farming is resource-efficient, sustainable, and profit focused, and it is necessary if we are to feed the huge global human population.[8] It represents a business strategy to conduct agriculture by means of the latest digital technology, including big data, artificial intelligence, and algorithms that track, monitor, and analyze. In other words, smart farming makes a great case study for scholars of surveillance. But with a few exceptions, researchers in visual and cultural studies have barely considered the role of drones in agriculture.[9] It is high time to discuss smart farming, the role of the drone, and its sensing of the earth.

Drones play a decisive role in smart farming. According to Brad Bolman, in October and November 2014 alone there were more than two hundred published news articles about agricultural drones, and these articles often celebrated drones as a new agricultural revolution.[10] Drones are seen as accelerators of smart farming since they can spray pesticides, monitor crops, and identify weeds. On the commercial website Postscapes, a platform for companies that deal in smart farming equipment, the benefits of the agricultural drone are summed up as follows:

> Drones are an affordable investment when compared to most farm equipment. They can pay for themselves and start saving money within a single growing season. By generating accurate field data you can:
>
> - Get much higher resolution data (up to 16× than traditional satellite methods) of your crop conditions.
> - Get a head's up on stressed areas, pest infestation, or if you have an irrigation leak anywhere.
> - Get a true count of plant growth so you can purchase insurance, plan labor schedules and predict yields.[11]

The sensorium of the agricultural drone gives information to the farmer, which in turn increases productivity and efficiency. The drone can deliver aerial images that register patterns of pest infestation, hydration, and irrigation; it can also collect images over time and monitor fields over long periods. But it is not only its aerial images that make the drone attractive for farming. Its capacity to become a data interface enables it to gather big data, and this is supposed to make farming more cost-effective, large-scale, calculable, and predictable. Drones are a key technology that fosters what Gabrys calls the "programmability of the environment,"[12] through which sensing technologies generate environments and configure new "techno-geographies."[13]

Many agricultural drones are equipped with multispectral imaging technology. In general, spectral imaging uses multiple bands of the electromagnetic spectrum. The human eye (and nonspectral cameras) can see visible light in three wavelengths (red, green, and blue), but spectral images can show more bands of the spectrum. They can thus make visible things that are usually invisible to the human eye. A multispectral image can capture wavelengths across the electromagnetic spectrum, including beyond the visible range (that is, infrared and ultraviolet).[14] In the military context, these forms of imaging are also referred to as thermal imaging methods, and Lisa Parks has analyzed them in connection with drone warfare.[15] Because this type of imaging can also register temperature, it is not an exclusively optical sensing technology. It is also thermal and kinetic, as it senses more than optically; it "feels" body temperatures and movements. Closely related to multispectral imaging is hyperspectral imaging, which uses "continuous and contiguous ranges of wavelengths (e.g., 400–1100 nm in steps of 1 nm) whilst multiband imaging uses a subset of targeted wavelengths at chosen locations (e.g., 400–1100 nm in steps of 20 nm)."[16] Hyperspectral images can create visibility in more nuanced ways. Besides military surveillance, they are also used in geology, mineralogy, eye care, food processing, and agriculture.

Multi- or hyperspectral sensing technologies embody a volumetric dimension insofar as the sensing apparatus that produces the images can go deeply into the earth or plants, revealing layers, textures, and densities that would otherwise remain unseen. These images add further wavelengths to the visible, expanding and inflating the space for even more layers. In particular, in conjunction with unmanned aerial vehicle (UAV) technology and farming, hyperspectral images can monitor the health of crops and soil. They can measure the chlorophyll concentration of plants: the less chlorophyll is detected, the more seriously a plant is infected. Consider the example of a hyperspectral image made by a drone to measure levels of fire blight infection in a pear orchard as part of a study conducted in 2020.[17] The UAV carried a hyperspectral Cosi camera, and its data were used to assess the wavebands that contained information about fire blight infection. The visual scoring ranged from the healthy (green) to the severely infected (red).

The UAV image in the pear orchard study serves as an extension of the human gaze—in this case, the gaze of the farmer who wants to save the orchard from infection. The technology enhances the farmer's capacity to discriminate between healthy and unhealthy crops, an activity that has been part of farming from time immemorial. Hyperspectral imaging broadens the farmer's visual range in unprecedented ways by giving it a volumetric dimension. This dimension is created by adding new distinctions to the objects in the image. In this case, we not only see plants and soil, but we can also discern which plants are infected and which are not, demarcated by differently colored layers. The image generates spaces of distinction between the healthy and the unhealthy, between plants with normal and abnormal chlorophyll levels, and between good and bad soil. Thus the drone's volumetric sensing makes visible layers of the materiality of the ground and the plants' textures that would otherwise be hidden from the human eye.

How can we interpret this multispectral drone gaze toward the earth within the framework of the Anthropocene? What attitude and affectivity does it express? And in what ways does it entail a vision of communities? Although I am more interested in demonstrating the vitalist potential of earth drones, it remains important to highlight their destructive force too. The drone's volumetric sensing embodies a military gaze onto the earth, and the dronification of agriculture reveals the normalization of military drone technology in civilian society. It is crucial to acknowledge earth drones' intrinsic militarist rhetoric and optics if we want to understand how these drones generate our environments and our relationships to the planet. UAV farming follows the logic of precision warfare, as the remote technology enables surgical operations in relation to crop management, pesticides, and infestations. Multispectral images of crop fields (in the earlier example, a pear orchard) are indeed reminiscent of battlefield images captured by remote warfare technologies.

In order to grasp this proximity between the farmer and the remote sensing warrior, let me give another example from scholarship on UAV multispectral imaging, this time the observation of "invasive" plants. There is a trend to use commercial drones' sensing technology to detect plants that are nonnative or alien to local ecosystems, as such species often cause economic and environmental harm. For example, a recent study on the spread of *Hakea sericea* (a bushy shrub) in Portugal exemplifies how the rhetoric of precision warfare penetrates scholarly discourse about drones and land monitoring.[18] The aim of the study is "to identify and quantify the areas covered by the invasive plant *Hakea sericea* using high spatial resolution images obtained from aerial platforms."[19] Note the language in the following passage: "*Hakea sericea* (Schrad and Wendl) is an invasive shrub which is colonizing large areas in the North of Portugal, and has been added to the list of invasive alien plants that was developed by the European and Mediterranean Plant Protection Organization (EPPO) in 2012. . . . Where it invades, Hakea sericea forms dense impenetrable stands and causes several negative impacts, including the reduction in indigenous species diversity and increases in biomass, fuel loads and the intensity of wildfires."[20]

Invasive plants can be harmful to biodiversity and cause severe damage to ecosystems (through fire, or by annihilating existing plants). Few engineers and biologists would even dream of comparing their way of speaking about plants to the rhetoric of UAV warfare. Nonetheless, the visual aesthetics of earth drones' multispectral images, and the choice of words used to describe them, are of critical interest, since they involve more than just "neutral" technical scientific language. The earth drones, their spectral imaging software, and the corresponding rhetoric in scholarly discourse mediate an attitude toward the earth that Grégoire Chamayou has dubbed the *necroethics* of the military drone.[21] This refers to the idea that the drone should be used to protect lives in the name of self-preservation: "While ethics is classically defined as a doctrine of living well and dying well, necroethics take the form of a doctrine of *killing* well. Necroethics holds forth on the procedures of homicide and turns them into the objects of a complacent moral evaluation."[22] Chamayou problematizes this view of the drone as an ethical and morally just weapon, since it breaks international law and hunts, kills, and exterminates human beings: it is literally above the law. Drones and their agents make decisions regarding whose lives are vulnerable and should be protected, and whose lives fall outside that frame and should be terminated.[23] Thus military drones and their sensorium envision human communities on the basis of a dichotomy between friend and foe. Such communities rest on the acquisition of a political identity, which in turn means that community members must preserve the group by fighting those who do not belong. In other words, communities are built on warlike forms of human interaction. The military drone enforces this view of community, as it clearly assumes antagonism and conflict a priori: it preempts violence by executing violence.

Of course, pear orchards and fire blight bacteria are not the same as human targets. Fighting a bacterium is not fighting the "war on terror." However, the connection between sensing and community-shaping echoes the military drone. The modes of sensing the earth, the scanning, and the sorting between good and bad, infected and noninfected, vulnerable and nonvulnerable, are similar. Like enemies, some plants are seen as threatening, invasive, and alien. Native and nonnative plants do not get along. Bacteria have to be exterminated in order to save healthy plants from the unhealthy. The life-taking optics of the military drone resonate in the life-making optics of the agricultural drone. As Bolman notes, "Precision *agriculture* is not simply an offshoot of precision *weapons*, but a co-emergent technological and cultural force, drawing on capacities and networks of knowledge with heavily military origins."[24] Similarly to the military drone, the pesticide drone suggests a vision of community based on exclusion, targeting, and elimination. Certainly, the pesticide drone is supposed to benefit the human community and help to solve the global food problem. But this optics of extermination also characterizes an attitude toward the earth that is based on instrumentalization and profit-maximization.

Although agricultural drones are supposed to have a vitalist agenda (delivering food, increasing sustainability), one has to ask for whom they work. For whom do they spray pesticides on infected spots? Who earns from the crops whose quantities are increased due to drones' big-data collection, and who gets to eat them? Who produces the drone technology and profits from it, and who owns it? Agricultural big business changes profoundly when drones enter the field. The farmer becomes a digital speculator, a data laborer, and a software engineer. If a farmer does not have the technical know-how or the capital to invest in these technologies, their chances of surviving the competition are low.

I have no intention to argue for a nostalgic vision of preindustrial farming. My aim is to show that agricultural drones sense volumetrically, and in doing so to uncover the family resemblance between military and agricultural drone sensoria and their visions of community. This kinship entails a specific attitude toward the earth, one that considers technology to be the best way to solve problems. The planet needs food, ergo we need drones and smart agriculture. This technological solutionism rarely asks long-term, big-picture questions about the systemic reasons behind food scarcity or the emergence of invasive plants. Agricultural drones see the earth as a patient that needs surgical precision therapy. The earth becomes a test site, a laboratory for maximizing profit and exploitation. Although the agricultural drone can come across as humanitarian, helpful, and climate-friendly, it can be a facilitator of humans' power in the Anthropocene and their hegemonic attitude toward the environment. The next section of this chapter discusses an artist who critically reflects on this by portraying megafarming and industrial agriculture.

Megafarming and the Drone as an Aesthetic Medium

Over the past decade, art photographers and multimedia artists have worked with new visual sensing technologies and representations of the earth. They often engage with discourse about the Anthropocene, reflecting on the human-centeredness of our geological age and how it might be negotiated in aesthetic works.[25] One well-known example is Edward Burtynsky and his large-format photographs of industrial landscapes, such as quarries, mines, oilfields, and shipyards.[26] However, I am interested in the drone as an aesthetic medium and how artists use it to reflect on agriculture, megafarming, and automated crop-raising, and for this purpose George Steinmetz's aerial photographs of twenty-first-century agriculture provide an excellent example.[27] Steinmetz works with drones, and he is invested in questions about the Anthropocene. He has long engaged with aerial photography, and indeed he became famous as a "flying photographer," shooting images from a paraglider or lightweight airplane. Recently he has also used drones to capture the view from the sky: "As an aerial specialist, I'm excited by the visual possibilities that drones have made available. I have a small squadron of quadcopters, each with different sizes and capabilities. But in spite of all the hype, drones have not entirely replaced the usefulness of airplanes, helicopters, or, in my case, motorized paragliders. They are only the shiniest new tool in the drawer."[28] Although Steinmetz emphasizes that the drone is just another tool, his drone images of megafarming exert an intense aesthetic power.

Steinmetz's visual chronicle *The Human Planet: Earth at the Dawn of the Anthropocene* takes stock of his recent work and is devoted to the representation of the different landscapes of the earth's continents.[29] Steinmetz states that although he did not understand himself as an environmentalist at the start of his photographic career, it subsequently became clear to him "that we are entering an era of limits, because we can't keep consuming resources at today's pace if we wish to leave a habitable planet to the next generations. The classic narrative of man versus nature might need to be rethought, as a narrative of man with nature."[30] According to Steinmetz, we need to change our perspective: we no longer live in the Holocene (the current geological epoch after the last Glacial Period) but are at the dawn of the Anthropocene. By the latter Steinmetz means a geological age when humans' power to dominate the globe has undermined the living space, the atmosphere, and the environment. Steinmetz portrays this process in his aerial images, tracing humanity's footprints in the natural spaces of the earth in our quests to grow food, generate energy, and build homes. A part of the project focuses on how humans harvest the biosphere, exploiting the land, seas, and skies to sustain our needs. The section of the book called "Harvesting the Biosphere" contains aerial shots of twenty-first-century agriculture in the form of factory farming. There are large aerial overviews of greenhouses in the Netherlands, cattle megafarms in Texas, cranberry pods in Massachusetts, and farming terraces in China. Although Steinmetz also used helicopters and paragliders, many of the images were shot by drone, and it is on these that I will focus here. All of the drone images have a supersized aspect, portraying the gigantism of modern megafarming.

Figure 5.1
Center-pivot irrigation systems on the edge of the Kubuqi Desert. Image by George Steinmetz. © George Steinmetz.

For example, consider a drone image that shows center-pivot irrigation systems on the edge of the Kubuqi Desert. The image has a documentary feel: it could be placed in a magazine article about pivot irrigation systems or waterwheels, or in a commercial advertisement for such technology. The image could be used to relate the narrative of how the desert was conquered and turned into farmland: how humans made the desert green, and how large-scale farming is beneficial and necessary to feed the planet's inhabitants. However, images never work without texts, discourses, and contexts, and the narrative of this image is in fact very different insofar as it appears precisely in *The Human Planet*. Being placed here, it tells quite another story about the earth, one that that has been left out of discussions of the "good drone" in agricultural contexts: it shows the earth as a space for profit maximation and industrialization. It shows the gigantism of megafarming.

Studying the three pivots in the image from the perspective of cultural criticism, we can observe that they evoke an effect of estrangement. We have the impression of looking at something alien and strange, something that is beyond comprehension. This effect manifests because the drone view from the sky intertwines distance and verticality with a closer view (in comparison with helicopter images, drone images are shot from lower altitudes). In doing so, the image captures a volumetric dimension. The volumetric here is not linked to the layered vision of the earth's surface that we encounter with agricultural drones. Instead, the volumetric acquires an atmospheric dimension, in the original sense of the word *atmos* meaning *vapor*. The drone image of the pivots catches the hazy cloud of vapor, mist, and dust, which in turn highlights the presence of the atmosphere and its importance for the earth.

A key thinker about the earth as an atmospheric space is Peter Sloterdijk, who writes about space as an interconnection between human and nonhuman materialities, a spatial form of being in the world. The metaphors of spheres, bubbles, and foam that Sloterdijk uses in his philosophy capture his idea of space as interconnected atmospheres.[31] This idea of interconnected space speaks to my desire to expand the sensorial palette of the earth drone, as it allows me to describe volumetric modalities of aerial space formations that go beyond the scopic regime. In Steinmetz's image of the pivots, the drone conveys a sense of the atmosphere. The veil of vapor and haze not only demonstrates the importance of water to life but also renders the image somewhat mystical, estranged, and incomprehensible.

This estrangement effect is heightened by the gigantism portrayed in the image. The pivots in the image are examples of what Morton has termed *hyperobjects*: objects that are massively distributed in time and space, such as the atomic bomb, or mega oil fields.[32] The pivots suggest a dimension of agriculture that seems too big to comprehend. Morton mentions five characteristics of hyperobjects: viscosity, moltenness, nonlocality, multidimensionality, and interobjectivity. In Steinmetz's image of the pivots, it is the characteristics associated with space that dominate—that is, massivity and nonlocality—since the image conveys a gigantism of crops and resources. The pivots might also appear to be a strange sign language that only extraterrestrials could understand; perhaps they are traces left behind by aliens visiting the earth.

The remote sensing technology of the drone and its photographer thus make the earth appear incomprehensible. They seem otherworldly, not humanmade. In such estrangement effects—as in the alienation effect found in Brechtian critical theater—Morton sees a pivotal moment that can change our thinking about the earth. Estrangement constitutes a step toward seeing the earth from a new perspective, beyond our common attitudes of exploitation and domination: "Seeing the earth from space is the beginning of ecological thinking. The first aeronauts, balloon pilots immediately saw the Earth as an alien world. Seeing yourself from another point of view is the beginning of ethics and politics."[33] This quote not only connects to my discussion of balloonists and the planetary in chapter 4 but also makes the drone into a medium that can generate an alternative view of the earth. The drone can portray agricultural megaprojects as hyperobjects, as things beyond our scale of comprehension, and this enables us to reevaluate our standpoints, views, and habits of looking. We learn from the drone about the earth.

The drone view of the pivots reveals the exploitation of the earth and evokes its vulnerability. Steinmetz's image highlights the earth's status as a test site for sheer profit interests. Thus it also makes a statement about communities. The image of the pivots (as well as Steinmetz's other drone images of agriculture) is completely devoid of humans. The human individual has disappeared, and the empty landscape merges into an abstract pattern or shape. The images do not highlight an individual human subject that cares for the earth; rather, this is an earth regulated by giant machines and smart technologies. One might wonder who steers these pivots. Where are the people that work here? Who buys this technology, and who eats the crops? The image does not answer these questions, but it makes us speculate about the impact of pivots on desert communities. In this way, the image of the pivots critically reflects a vision of community that is devoid of individuals and where technology has taken over from human labor. Moreover, the community aspect is also pertinent to the idea of a specific attitude toward the earth. Steinmetz's drone images register how megafarming expresses a certain attitude toward the planet, one that is based on exploitation, profit, and the depletion of natural resources. In other words, the human and the earth shape a form of community in which technology instrumentalizes the earth to make money through large technological investments. Thus Steinmetz's images show the earth's vanishing aesthetics, and even a peculiar beauty of destruction. The next section of this chapter further traces the drone's capacity to observe, register, and negotiate the earth in the Anthropocene. Again, the idea of volumetric sensing will be vital, as I will speak about artworks that open up the flattened and scopic paradigm, deeply sensing the earth beneath the surface and through geological time.

The Deep Sensing of Earth's History: The Drone and Ecologies of Duration

Civilian drones frequently occur in geological research, where they provide information about changes in glacier formations, water streams, erosion patterns, and wildlife conservation.[34] J. D. Schnepf has called such drones *eco drones*, raising a critical discussion about animal surveillance and the humanitarian drone's military entwinements.[35] Although I remain cognizant of ingressive militarism, this section of the chapter discusses how landscape designers and artists can free earth or eco drones from military surveillance contexts and experiment with them as tools with which to witness the planet's environmental challenges at the dawn of the Anthropocene.

The visual artists and scholars Michele Barker and Anna Munster are creating the ongoing visual installation *Ecologies of Duration*, which connects drone sensing, geological time, and the observation of the earth.[36] They have compiled up to twenty "ecologies," that is, montages that focus on a specific geolocation sensed by a drone—for example, in Kilpisjärvi, Finland, in 2019.[37] The drone in this installation senses the earth in a synesthetic fashion, combining sounds, images, and touch. The drone's sonic dimension is crucial because the installation plays a variety of earth sounds, such as dripping water or the sound of wind. One also hears the soft, deep humming of the drone, and this simulates the experience of listening deep into the earth with geological sensing technologies, as the soundscape gives an impression of geoformations and material structures. These soundscapes do not distance us from the earth, but bring the geoformations closer to us.

This feeling of proximity to the earth is also conveyed at the installation's visual level. The drone camera can zoom in to and recede from surfaces, and it registers the flow of water and the light over the lake. It also films mist and fog in places that are hard to see by human eyesight alone and would remain obscure without a drone. The presentation screen for *Ecologies of Duration* is doubled, and thus the visual installation suggests the aesthetics of a montage, as in aesthetic works by Harun Farocki, or in Ignacio Acosta's use of double monitoring (see chapter 4). The montage's multiperspectivalism undermines the drone's vertical scopic regime, which is enhanced by the filming technique of hovering and the drone's Steadicam-like view of the earth. This Steadicam view is somewhat unusual for drone films: drones are more associated with flying in circles over us, or following a target. The Steadicam view here is not distanced, but close and attached to the ground. The artists describe this effect as a type of "terrain hugging" in

Figure 5.2
Michele Barker and Anna Munster, *Ecologies of Duration* (2020). © Michele Barker and Anna Munster.

which the drone seems to caress and intimately touch the earth: "In working against the scopic regime of the distant, aerial view, we try to work with drones . . . as a machine 'agent' capable not only of tracking and targeting but potentially also of gesturing, in the sense that Erin Manning proposes as 'reaching-toward.' Manning describes touch as a relational encounter that alters and displaces spatialities and temporalities."[38] This reaching toward the earth integrates touch into the sensorium, thereby counteracting the violent military gaze of the drone's scopic regime. The drone conveys the idea of caring for the earth, a form of ecological care in the sense discussed by Maria Puig de la Bellacasa. Puig de la Bellacasa suggests a model that decenters the human subject in more-than-human webs of care, with the potential to reorganize human–nonhuman relations and communities into nonexploitative forms of coexistence.[39] The drone in Munster and Barker's work is a medium of this kind of care because it connects the earth and humans in a community via noninvasive and noninstrumental sensing. Again, the dimension of community here is based on a relation and interaction between human subjects (and their sensing technology, the drone) and the earth. Similarly to the community in the artwork *Cloud Face* (discussed in chapter 3), this community is defined by a specific way in which the human relates to the earth. In contrast to the technical images of megafarming discussed in the previous section, the drone in this case represents a technology that enables an embracing and noninstrumentalized mode of community-building.

In other words, in *Ecologies of Duration* the drone is a sensitive and affective witness to changes in the earth's surface and atmosphere. Its sensing, rather than being destructive, is conserving and protective. Its synesthetic sensoriality—that is, its rhythmic, sonic, visual, and haptic embrace of the landscape—embodies a volumetric sensing that simulates a reach into the depths of the earth but also connects closely to its surface and its atmosphere. The drone sensorium creates a volumetric interface of earth experience as it sensorially comprises the earth from multiple directions: from the ground to the sky, from within, from a hovering view, and from below. The drone-witness registers the impact of anthropocentric power on the earth. It makes visible the continuity of geological time and suggests a reflection on the relative shortness of the human era in the earth's history. The viewer of *Ecologies of Duration* starts to listen to the earth's sounds as reverberations of its geological ancientness. The surface images in the drone's film clips account for the long time periods over which geological formations emerge; fog and mist over the water speak to the precariousness of the atmosphere. The earth and the drone together tell a story of the *longue durée* of the earth's evolution, thus providing a counterpoint to the accelerated destruction that human industrialization has wreaked on the planet. *Ecologies of Duration* also reveals a community aspect, that is, a noninstrumental and nonexploitative mode of relating and interacting with the earth, which in turn provides a generative imaginary matrix to build more sustainable and planetary forms of communities. In the next part of this chapter, the drone's aesthetic power to reflect the Anthropocene will remain in focus as I discuss aesthetic projects that use drones to imagine and even build sustainable communities.

The Volumetric Sensing of the Drone in Landscape Design

Drones have become a key tool in landscape architecture, a field that has always had an interest in new technologies to enable mapping, surveying, and spatial data gathering.[40] One tradition in landscape architecture and its technologies of measuring can be characterized as "grounded": the surveyor stands on the ground with a technical measuring apparatus, such as a theodolite, an optical instrument developed during the eighteenth century to measure angles on horizontal and vertical planes. In the theodolite tradition of landscape measuring, the surveyor conducts a precise triangulation of space, dissecting and abstracting the site according to specific geometrical features. Another approach is aerial landscape measuring, surveying the land from the sky, which began when surveyors observed the land from hilltops, church towers, kites, balloons, and, eventually, airplanes. The aerial measurement technique provided views of the earth from above, suggesting its continuity, interconnectedness, and wholeness.[41] The advent of satellite images of the earth in the late 1950s marked the apex of aerial landscape surveying. Sputnik 1 in 1957, and the images of the earth as a blue marble taken from Apollo 17 in 1972, were the beginnings of satellite imaging, which fundamentally changed our cultural assumptions about virtual and physical space and our modes of representing space.[42] Satellite images show the earth's landscape as urbanized, and digital mapping gives multilayered information about topography, energy levels, population densities, and meteorology.[43] In addition, satellite images have triggered a democratization of aerial views. While aerial shots of landscapes from planes and helicopters are expensive, satellite images are relatively easy to access. Today, programs such as Google Earth provide an aerial view of the planet to anyone with access to an Internet browser.

In recent years, researchers have acknowledged this democratization of aerial views of landscapes via satellites. However, some scholars also note the limitations of this technology for landscape architecture. Karl Kullmann, for example, remarks, "While this lofty position reveals cultural and natural patterns and associations on the ground, the nuances and details that enrich the landscape are often camouflaged from view in shadowed, interstitial and underneath spaces."[44] In addition to problems with image resolution, there is the fact that the surveyor is remote and detached from the ground, and thus is not in touch with the affective materiality of the landscape. For example, Brett Milligan notes the "indirect and detached, or remote access to the landscape medium"[45] that is provided by satellites.

Although the drone is an aerial surveying technology, it can counteract this detachedness and resituate the landscape architect on the ground. It does so not by virtue of its vertical features, but rather through a form of volumetric sensing.[46] In landscape surveying, drones have become not only a new technological object but also a technological sensorium that facilitates the "aesthetic and techno-experiential interpretation of sites."[47] Drones are taken out of our societal preoccupations with surveillance and warfare and used instead as tools that can co-construct "qualitative relationships among technology, humans, and landscapes."[48] The drone can change the relationship to the landscape, making it more immersive, dynamic, and intimate by embedding its pilot in the sensual experience of the surveying process: "The act of launching the drone upwards from the ground reverses the downward zoom of satellite imagery and places the landscape architect *physically* on the site and *virtually* within the frame of the map."[49] As Kullmann highlights, this involves an inversion of perspective, no longer from above down to the ground but from the ground up into the sky. This look upward situates the landscape surveyor on the ground as well as suggesting a volumetric sensing of the sky above. The drone reaches out from below and models atmospheric space into the sky. It can do this not only by tilting an immersive first-person view (see chapter 2) but also by using software that transforms drone images into three-dimensional images.

The work of Christoph Girot and his research team at the Swiss Federal Institute of Technology in Zurich on digital and robotic landscapes has been groundbreaking here. Among many other projects, Girot's "school" of landscape architecture focuses on new digital technologies and their power to fabricate, model, and visualize landscapes. Drones play an important role in Girot's work because they are used to scan landscapes from the air and collect data about size, surface, and topographical changes.[50] However, in addition to filming and monitoring landscapes, the drones collect data that are used to form the basis for new software usages. The drones' geospecific data are used in point cloud models, sets of data points used to model and shape a three-dimensional animation by means of dedicated software. This software is called lidar, and it involves the method of measuring distances by illuminating the target with a laser beam and measuring the reflection with a sensor. Differences in laser return times and wavelengths can then be used to make three-dimensional digital representations of the target. The final point cloud model produced by the lidar scanning of a landscape can thus suggest a three-dimensional topography, simulating changes over time and showing the past, present, and possible futures.[51] This method is also linked to photogrammetry, a technology to obtain topographical data by measuring photographic images, which involves putting two-dimensional images together in a kind of mosaic to suggest three-dimensional space. Photogrammetry was practiced as early as the nineteenth century, for example by the French surveyor and photographer François Arago, who used it to make topographical maps. The term *photogrammetry* was coined by the architect Albrecht Meydenbauer, who published an article about this technology in 1867.[52]

In this chapter, I am specifically interested in a three-dimensional form of photogrammetry in combination with the drone: the extraction of three-dimensional measurements from two-dimensional data. This is precisely what drones can do for landscape architecture by implementing point cloud models. In Girot's approach, the point cloud model opens up a sensorial experience of a physical place. The drones, not being bound to a vertical paradigm of seeing, can be used to project three-dimensional visions of landscapes and territories. The drone is no longer an aggressive eye in the sky; it volumetrically senses alterations in the landscape due to climate change and provides a potential trajectory for sustainable landscape planning. The next section traces this photogrammetric volumetric sensing of the drone in works by contemporary artists. The experimental multimedia artist and researcher Keith Armstrong, who works with new media technologies, site-specific electronic arts, public arts, and art–science collaborations, is of particular interest here.[53] His aesthetic projects utilize the drone and its generation of three-dimensional topographies to project future imaginaries of communities. In contrast to Girot, Armstrong not only uses the drone as an aesthetic medium to model and observe landscapes, he also critically reflects on drone-generated landscapes as a political intervention in light of the Anthropocene.[54]

Drone Art, Volumetric Sensing, and Postnatural Communities

Armstrong's works focus on social and ecological justice, sustainability, and the future of this planet, drawing his audience into experimental processual practices that entwine digital media technologies, science, philosophy, and ecology. His artworks negotiate with the position of humankind on earth by exposing the fragility of our environment and drawing attention to the "unstoppable catastrophe in the biosphere, cryosphere, and atmosphere."[55] Among other digital technologies, Armstrong uses commercial drones in his art-science and community projects. Hence his work is more than relevant to my inquiry into the drone and its role in observing the earth during the Anthropocene.

Armstrong seeks to remove drones from military and economic-industrial contexts: he "utilizes drone technology in order to invert its associations with surveillance and warfare into something much more life-affirming, seeing these aerial robots as tools for transforming our sense of ourselves within interdependent planetary ecologies."[56] His art projects transform drones from tactical into ecophilosophical media. For Armstrong, *ecosophy* is a "series of ecologically oriented principles, or a form of personal practice"[57] that explores the tensions between human and nonhuman in light of environmental decline, climate change, and species extinction. An important notion here is *environment*, which Armstrong, echoing Gabrys, conceptualizes as an interwoven, mutually constitutive relationship between humans and nature, rather than as an entity that is separate from the subject.[58] In this respect, Armstrong understands technology as a key generator of human–environment entanglements; in his ecosophy, sensor technologies constitute the earth, which can even become a sensor itself.

This expanded view of the environment can be meditated, activated, and negotiated by art. For Armstrong, art is a vital channel through which the artist, the audience, and the scientist can interact affectively with the materiality of the environment. Art's capacity for deep investigation, non-linear insight, and affective power can offer new creative perspectives on the earth. Art represents a critical discourse but also offers possibilities for concrete interventions to change how communities live together. This activist aspect of art is fundamental to Armstrong's community projects, *Future_Future?*, *Seven Stage Futures*, *Re-Future*, and *Change Agent*.[59] In these works, Armstrong and his team of artists visit various communities and settlements in South Africa and actively engage in community projects with local people. All of these projects intersect with Armstrong's initiative to build communities in South Africa, which began in 2014. Thus "community" takes on a very concrete and real dimension, as artists and locals build and design houses and infrastructures together. The guiding idea is "to design and build low-cost sustainable build-ings and models for living with residents, scientists and international devel-opment organizations in South Africa's informal settlements, working with international development workers, architects, and sustainability scientists. Working on the ground with local residents I then assisted them to re-image and re-present that work for new audiences to build understanding, support and critical mass around the work, including a presentation at a series of community festivals."[60] The communities around these projects emphasize the local, but they do not pursue a nationalist agenda. Instead, they are more fluid collectives, which emerge from the projects, the communities that already existed in these locations, and the temporary communities formed with the artists, documenters, and audiences.

Drones play an important role in this community-building and remodel-ing process. They document the building developments, and they fly over the settlements to gather data about village structure, infrastructure, and road conditions. The drone camera is not always an eye in the sky; it also films settlements' niches and in-between spaces, suggesting a volumetric grasp. Indeed, an additional step in the work further enhances this volu-metric aspect: the drone images are processed by photogrammetry software that maps them in three dimensions. Thus, similarly to Girot's landscape architecture work, the commercial drone that flies around the South Afri-can village gathers two-dimensional images that form the basis of three-dimensional visualizations of the landscape.[61]

However, Armstrong does not just display these three-dimensional visu-alizations at local art exhibits. Working in close cooperation with local people, his team manipulates and changes the three-dimensional images into projections showing how the village might look in the future. This future aspect comes into play because the software that processes the drone images projects new plans and new models for houses and infrastructures. Here again, the drone follows a volumetric sensing process; together with the human agents, the drone builds visions of communities on the ground. The software can then design building infrastructures and model how they might appear. Thus this community vision does not grow out of vertical surveying; rather, the three-dimensional drone images tilt the ground and turn the perspective upside down. The drone database provides an experimental and creative matrix for the imagining and remodeling of the village. But Armstrong's work does not stop there: the local community and artists actually build some of these creative three-dimensional models, constructing what Armstrong calls *postnatural houses* out of sustainable, low-cost materials, such as clay, recycled glass, and tires.

This image shows an example of a postnatural house in the village of Caleb Motshabi in South Africa. For this house, the artists and locals used natural building methods (mud bricks) and integrated recycled materials (glass bottles). Once a house has been remodeled, its images are shared at festivals, reaching new audiences and communities. The festivals are organized as community-led Meraka, which Anita Venter defines as follows:

Figure 5.3
Seven Stage Futures (2017). Preparation for the Meraka: aerial view of Mokoena and Ellen's residence, Caleb Motshabi, South Africa. © Photographer: iFlair.

"Meraka can be described as a cultural place where different generations come together and skills transfer is done between the elders and the youths. This research (the Meraka events) [is] a constructive attempt to turn theory into practice in order to challenge post-apartheid building codes that are biased to devalue centuries of indigenous building knowledge systems, and that hamper the potential that innovative building has in advancing pro-poor community livelihood futures."[62] The integration of the building projects and drone technology into the Meraka festivals embeds Armstrong's projects in the local culture. This is important, as Armstrong's work might all too easily mirror a colonialist gaze—a white artist coming to South Africa and wanting to build communities, improve infrastructures, and offer Western technology (that is, drones). Indeed, some researchers have accused drone projects in Kigali, Rwanda, of adopting a colonialist gaze since Africa is seen as a continent in need of help from Western technologies and infrastructures (see chapter 7). But Armstrong's integration of his community art projects with local institutions, people, and customs (such as the Meraka) works against this colonialist perspective: the projects are mostly administered, controlled, and determined by local people. The drone in Armstrong's projects is only one medium (among many others, including exhibits, building processes, festivals, performances, catalogues, and websites) to articulate his ecophilosophical ideas. Nevertheless, the drone is a key tool for imagining future communities. Instead of killing and destroying, the drone creates visions of sustainable living. In close collaboration with human agents, its volumetric sensing fosters community models built on giving back, sharing, recycling, and restoring.[63]

Postcarbon Communities

This chapter shows that the drone can be a quintessential medium of the Anthropocene.[64] Drones can originate, facilitate, foster, accelerate, and amplify the Anthropocene's destructive energy. But drones as aesthetic assemblages can also reflect critically on these processes and perform interventions. Their ability to sense volumetrically—to sense spaces up from the ground into the sky, below the ground into the earth, and in-between layers, niches, volumes, and voids on the surface—brings drones into contact with the Anthropocene. These volumetric sensing modes of the drone embody a fertile ground for imaginaries of communities between humans, drones, and the earth environment. As I show in my discussion of megafarming, the drone can exemplify exploitative attitudes toward the earth, although it is often couched as humanitarian and sustainable technology. In smart farming, the drone is eagerly advertised as a technofix for world hunger, but it generates communities that recall the dichotomous military optic of friend versus foe. However, as this chapter highlights, this toxic optic does not exhaust the earth drone's power to open critical perspectives, negotiations, and interventions in the Anthropocene. This power is particularly visible in experimental drone art that engages with volumetric earth sensing. The drone can witness the Anthroprocene and register, reveal, and observe the need for what Morton calls *postnatural ecology*, as I suggest in my analysis of Steinmetz's megafarming drone photographs. But the earth drone is not only a tool for documenting the Anthropocene. It is also a sensing machine that—together with its human agents—can generate alternative environments and communities. Earth drones in experimental aesthetic projects acquire the speculative power to imagine what Paul Cureton calls "post-carbon futures."[65] These imaginaries of postcarbon communities are visions of future collectives where humans, technology, and the earth live together in a more mutually sustainable and ecological relationship. Art projects by Munster, Barker, and Armstrong demonstrate this constructive potential of the drone. Postcarbon communities do not just point to the necessity of green and fossil-fuel-free energy, they also suggest a change in our thinking about our planet, its vulnerability, and the potential ending of the Anthropocene. These communities embody aesthetic attempts to create alternatives to the Anthropocene—or, following Bernard Stiegler, to create training grounds for the more hopeful "Neganthropocene"[66]—providing perspectives, possibilities, and potentials for a more sustainable future and moving beyond the often dead-end and fatalistic discussions about the end of humanity. The drone in these instances exercises a futurizing power in which it can become an agent for the social good, and for utopian dreams of more sustainable planetary communities. These visions, however, should not lead us to fetishize the drone or celebrate a posthumanist techno-utopia. Drones and their volumetric sensoria still have to be handled with care and reflection: they can integrate the human, but they can also expel it.

IV The Nonhuman

6   Swarm Sensing and Multitude Communities

Drones can fly in swarms. They can aggregate, move collectively as self-propelled entities, and even steer themselves via their own interfaces.[1] Although the deployment of military drone squadrons is still in its infancy, the armed forces are eagerly investing in research and development on drones, swarm robotics, and artificial intelligence.[2] And recent deployments of so-called Kamikaze drone swarms in the Ukraine have given us a glimpse of future warfare. In aesthetic imaginaries in literature, film, visual art, and computer games, drone swarms are often militarized. In the computer strategy game *Drone Swarm*, the player has to control a swarm of thirty-two thousand drones in a battle against a powerful alien fleet; in the film *Oblivion* (2013), drones are used to fight evil extraterrestrials. But drone swarms are also a hot topic in domestic realms too, such as environmental monitoring (detection of wildfires and pollution), policing (crowd-monitoring), and delivery solutions.[3] And accordingly there are aesthetic drone swarm imaginaries that do not have warfare or military settings, and these imaginaries are of interest to me in this chapter. Think, for example, of the drone display at the Tokyo Olympic Games in 2021; or the 2020 New Year celebration in Shanghai, when drone swarms flew above the Huang He river and sketched the shapes of Chinese characters, flowers, and a walking human;[4] or Cirque du Soleil's 2014 Broadway show, with dancing drones disguised as lampshades.[5] All of these could be seen as drone swarm formations.

Another example is the Hive, an architectural design created in 2016 by Hadeel Ayed Mohammad, Yifeng Zhao, and Chengda Zhu. The design was made for a competition that sought proposals for a new building at 423 Park Avenue in Manhattan. The plans for the Hive show a 1,400-foot tower that functions as a docking station for thousands of drones. The drones are supposed to fly in swarms around the tower, land on it, and receive directions for their next delivery tasks. The Hive's name illustrates the connection between drones and the entomological world: drones behave like bees, in hives and swarms, and they thus represent a nonhuman mode of sensing, communication, and mobility. The tower itself suggests a living surface: the platforms move back and forth, and so the tower's exterior constantly changes. A guiding idea behind the Hive was the proposed revision of the Federal Aviation Administration's air-zoning regulations through the creation of a vertical highway. Largely intended as a recharging station for commercial delivery drones, the Hive provided an alternative urban imaginary to the much-criticized "Millionaire's Matchstick" that currently stands at 432 Park Avenue.

How do civilian drone swarms sense and perceive their environment, and what kind of sensorium do they have? How are drone swarm sensoria reflected in the aesthetic realm (popular culture, literature, visual art)? What community visions do they suggest? Does the nonhuman, zoological element of the swarm necessarily lead to an inhuman vision of society? Or are there ways to think the drone swarm constructively and in relation to the human? In order to find answers to these questions, this chapter discusses

Figure 6.1
Hadeel Ayed Mohammad, Yifeng Zhao, and Chengda Zhu. The Hive: model for 423 Park Avenue, Manhattan (2016). © Hadeel Ayed Mohammad, Yifeng Zhao, and Chengda Zhu.

the sensorium of the swarm in three different aesthetic drone imaginaries from popular culture, modernist fiction, and protest art. I am interested in how artworks interpret drone swarm formations as a (partly) nonhuman form of sensorium, but also in what these artworks reveal about how drone swarms sense, and how this differs from human perception.

The first example is "Hated in the Nation," a 2016 episode of the popular Netflix series *Black Mirror*.[6] In this episode, artificial bee swarms have been introduced to counteract the decline of natural bees. The artificial bees are supposed to be a "sustainable" technological invention, although as we will see, the invention goes disastrously wrong. This imaginary and its dystopian implications for the human community are contrasted with my reading of Ernst Jünger's novel *The Glass Bees* (*Die gläsernen Bienen*, 1957),[7] whose swarms of robotic drone bees I will examine as a multitude. Finally, after having investigated these filmic and literary imaginaries, I will analyze the politics of drone swarming among protest communities of activists and artists. I argue that these three aesthetic imaginaries of drone swarms highlight the nonhuman aspects of the drone sensorium, and in doing so suggest visons of community that go beyond sociocentric definitions. The community visions generated by the drone sensorium of the swarm can go terribly wrong: they can be violent, inhumane, and destructive. But they can also evoke nonessentialist, posthuman, and fluid models of communities as multitudes. Before I trace this dialectic of the drone swarm, a short excursion into the concept of the swarm from cultural, media, and sensorial viewpoints is in order.

Cultures of Swarms

In scholarly debates during recent years, the swarm has become a popular trope for formations of social revolutions, emergent forms of digital intelligence, and the future of remote warfare. The units of a "swarm" can be animals (swarms of insects, shoals of fish, flocks of birds) or other living organisms (bacteria, plankton), and this zoological and biological model of the swarm can also be transposed into the realm of technology (robots, drones, algorithms), political organization (smart mobs), and even representations of knowledge (Wikipedia, blogs).[8] Swarms consist of self-controlled singular units that can form (often in a sudden dynamic move) an "intelligent" whole.[9] This whole, however, constitutes much more than the sum of its individual parts. Swarms follow a logic that is organized not by linear causality, hierarchy, or the principle of addition, but by a dynamic of emergence and self-control without a center. Alexander R. Galloway and Eugene Thacker describe the swarm as a network that has no single, centralized principle of power but is defined by a paradoxical simultaneity of control and emergence.[10] A swarm does not relativize power; it presents a contradictory togetherness of formlessness and strategy, amorphousness and coordination, decentralization and centralization. The swarm presents an epistemological paradox: "Humans thrive on network interaction (kin groups, clans, the social), yet the moments when the network logic takes over—in the mob or the swarm, in contagion or infection—are the moments that are most disorienting, and most threating to the human ego."[11] Thus swarms entail a nonhuman form of intelligence that must be understood not as a single "mastermind" but as a "hive mind," a form of collective thinking that is also a particular way of sensing. A hive mind senses synesthetically, combining visual, haptic, olfactory, kinetic, and acoustic sensors, and sharing these sensorial impulses simultaneously and collectively. Sensing is not bound to the individual but is continuously interconnected, shared, and networked within the swarm. These sensing processes are not bundled and steered by a central sensorium; rather, the sensorial impulses organize themselves through an emergent collective dynamic.

This nonhuman sensing and intelligence often appears deeply uncanny to humans: "Suddenly, the many pre-exist the One, animal packs operate without heads (without one specific reason or leader), and suggest logics of life that would seem uncanny if thought from the traditional subject/object point of view."[12] Technological swarms are no less uncanny than zoological swarms: drone swarms can give us the creeps. Citizens and authorities alike were frightened and puzzled by the mysterious appearance of drone swarms over the Colorado and Nebraska prairies in 2020. Many witnesses described these swarms as uncanny, as no one knew where the drones had come from or who was steering them.[13]

The realm of culture and aesthetics offers a privileged stage for uncanny imaginaries of swarms. Swarms often evoke the deep-rooted fear that they herald the end of humanity. Think of Salvador Dalí, who injected his terror of locusts into his surrealist paintings, playing off the ancient biblical connection between locusts and the coming apocalypse.[14] In popular culture, the animal swarm frequently appears in horror scenarios. Alfred Hitchcock's film *The Birds* (1963) impressively demonstrates how the change from a single bird to a whole flock can evoke sheer panic. In Frank Schätzing's novel *The Swarm* (*Der Schwarm*, 2004), a biological swarm intelligence at the bottom of the sea seeks to destroy human civilization in order to save the global ecosystem. In *The Invincible* (1964), a science fiction novel by Stanisław Lem, a human space mission finds intelligent behavior in a swarm of small particles. Swarms often embody the zoological and bestial "other" by mimicking the evolution of life beyond human control.[15] No artwork exemplifies this better than "Hated in the Nation."

Drone Swarms and Killer Bees

In "Hated in the Nation," the main protagonists are artificial bees known as autonomous drone insects (ADIs), designed by the company Granular.[16] They are the same size as natural bees; they have wings, insectoid legs, and eyes; they even buzz. In the episode, natural bees are dying en masse in the UK, and the artificial drone bees are intended to counteract the phenomenon of colony collapse disorder. This disorder is a real-life phenomenon that can lead to the extinction of whole bee colonies;[17] it is caused by environmental pollution, monocultures, and climate change.[18] The episode thus plays with the real-world threat that bees' extinction poses to biodiversity, ecosystems, and climate zones. One of the Granular engineers states that the disappearance of natural bees would cause an "environmental catastrophe,"[19] and that the development of ADIs as stand-ins is necessary to save the world. In other words, Granular nods toward the idea of the good drone, providing a technological solution to the dying of bees. But as the episode unfolds according to the bleak aesthetic that is typical of *Black Mirror*, these good drone bees bring death instead of life.

"Hated in the Nation" is conceived as a tragic murder mystery. Two detectives, Karin Parke and Blue Coulson, try to solve the unexplained deaths of people who have been exposed to social media "shitstorms": the journalist Jo Powers, who had received death threats (tweets saying "death to @JoePowerswriter") after writing a column about a disability rights activist's suicide; and the rapper Tusk, who had insulted a fan. Both Powers and Tusk suffered a cruel death when a small metal sensor penetrated their brains; as we discover, they were attacked by Granular-designed ADIs. The detectives visit Granular to gather more information about the automated bees. As soon as they enter the company's large foyer, they have their first encounter with an artificial swarm: thousands of ADIs form a hovering, three-dimensional G (for Granular). Coulson touches the G and disturbs the formation, but the bees reform it after just a few seconds. Parke comments: "Jesus, I didn't expect to find myself living in the future, but here I fucking well am."[20] The artificial bees' swarm formation puzzles the otherwise smart detectives, as it represents a collective organism beyond their comprehension. The swarm emanates epistemological ambivalence: it seems to be intelligent, but it has no visible control center that steers its decisions. After this encounter with the swarm, the detectives are led through Granular's greenhouses, where many artificial bees busily swarm about, pollinating blossoming orchards and building artificial hives. Granular engineer Rasmus Sjoberg explains to the detectives that the bees are solar powered and have visual sensors that work with pattern recognition.

These drone bee swarms thus have their own zoological-technological sensorium. They sense visually because they have cameras and pattern recognition, but due to their beelike way of flying, their vision is not scopic and vertical but multidirectional, versatile, and motile. Like a natural bee's faceted eye, they perceive their environment in mosaic-like, fluid, and acentric ways. Their sensing also has a kinetic-volumetric dimension (see chapter 5), since they can move fast, at many angles, and in different environments. Later, when we see the drone bees on their kill mission, it becomes clear that they sense according to a swarm logic, absorbing visual and datafied information that they share within the swarm in real time. Their buzzing provides a sonic dimension: although the constant humming is mostly presented as a source of terror to the humans that encounter them, it also adds to the swarm's own sensorium. When the ADIs are not on a kill mission, the buzzing vibration of their wings shakes pollen from flowers and triggers pollination. In other words, the ADIs are a prime example of the nonscopic multisensoriality of the drone (swarm).

As their name indicates, ADIs have attained a certain degree of autonomy. In a conversation with Parke, Sjoberg explains that the ADIs can navigate by themselves as datafied interfaces; in effect, they are computerized learning systems that can make themselves smarter.

Parke: *They* navigate? You don't . . . you know . . . steer them?

Coulson: They're autonomous. That's right. Isn't it? They make their own decision. They look after themselves.

Sjoberg: Yes. You see the ADIs cover the whole of the UK. We couldn't command each one individually ourselves. It is just not individually possible. No. We simply set the behavior and leave them to it. They even construct these hives themselves. They reproduce.

Parke: They reproduce?

Sjoberg: Yes. Each hive is a replication point.

Coulson: It is like a 3D printer, basically.

Sjoberg: Exactly. They create duplicates of themselves, create more hives and spread out exponentially.[21]

Granular has designed artificial bee swarms that are partly independent. They can program their own reproduction and their own locations (hives). The ADIs attain a swarm intelligence that differs from human intelligence. A swarm has no "head"; rather, as Galloway and Thacker note, the swarm is faceless,[22] an amorphous conglomerate that is nevertheless able to maneuver collective action (build hives, reproduce, pollinate). "Hated in the Nation" thus portrays a technological fantasy according to which nonhuman drone swarm intelligence will be beneficial for society and the environment. The artificial bee swarms and their hive mind not only embody a fusion of technology and nature but also have a constructive purpose: to save the planet and help natural bees. Technology is seen here in a utopian framework, since the robotic bees generate artificial life, reproduction, and evolution. However, as is to be expected from *Black Mirror*, this technofix fantasy of the bee swarm as the savior of humans and the planet goes completely haywire. Echoing Walter Benjamin's ideas about technology, "Hated in the Nation" suggests that society is not mature enough to handle technology and to use it in a constructive and nonviolent way. In fact quite the opposite is revealed: the robotic bees become a war machinery for mass killing. Indeed, we discover that Granular's ADI project is sponsored by the military, although the government's original intention was "only" to use the project for surveillance rather than assassination. The drone insects have become handy devices for the military to spy on citizens, detect data, and surveil every inch of daily life—the ADIs (four thousand per hive) are spread in a large network all over the UK.

As the episode unfolds, we learn that the drone bee swarms have fallen into the hands of a hacker who is on the rampage, seeking to avenge a friend he lost to suicide due to social media bullying. The hacker has connected the artificial bee swarms to the cruel and deadly "Game of Consequences," a kill game fostered and facilitated via social media. People who have posted negative comments, critiques, or hate speech are exterminated by the bee swarms, whose deadly stings can be activated through the hashtag #DeathTo. The journalist and rapper were only the first of many victims. Ultimately, the two detectives cannot deactivate the bee swarms, which go on to kill 387,036 people who have been targeted via the hashtag.

The figure of the hacker gives the faceless swarm a cruel face. When he appears, the swarm loses its enigmatic otherness. The hacker as master removes the swarm from its paradoxical simultaneity of decentered and centered power. The interesting symbiosis of zoological, human, and technological sensing, which supposedly had the potential to save the planet, is destroyed by the hacker (and the military's involvement in the ADI program). The hacker erases the otherness of the swarm by showing that it is controlled by his programming, hacking, and interference. By attaining a human control agent, the swarm of drone bees becomes utterly inhuman and destructive. The episode's plotline turns the technozoological formation of the drone swarm into a violent technology that falls into the hands of a murderous psychopath and is hijacked by the interests of the military.

Ultimately, the drone bee swarm imaginary in "Hated in the Nation" portrays the swarm as a sensorium that generates visions of inhuman communities. It is the sensorium of the killer bees, their sensorial versatility, their programmability, autonomy, and intelligence, that unleashes their destructive power on the human community. It is interesting that these social imaginaries are closely connected to communities engendered on social media via datafied networks and interfaces. The episode portrays social media communities as themselves built according to swarm logics: the tweets, likes, and messages emerge in unpredictable ways, exploding into numerous clicks that connect masses of users within seconds.[23] Thus the episode comments on the dangers of social media communities, which can be ruled by a violent, cruel, swarmlike mob. In doing so, it also takes a critical stance on autonomous surveillance and shows its dehumanizing effects by engaging with our ancient fears of insect swarms. The technology of the artificial bee swarm gets out of hand, annihilating instead of creating and continuing life. This aesthetic drone swarm imaginary also recalls the film *Slaughterbots* (discussed in chapter 3), in which a swarm of drones with facial recognition software kills politically active students.

But my key interest lies in drone swarm imaginaries where the otherness of the swarm is not militarized, and its uncanniness does not turn into horror. There are literary authors, artists, protest groups, and urban planners that experiment with drone swarm imaginaries in more utopian and life-affirming ways.[24] It is important to note that swarming can have vital connotations. The German word *schwärmen* foregrounds this beautifully: it can mean to flock or to sprawl, or to invade something in unpredictable and uncontrollable ways (*ausschwärmen*). *Schwärmen* has a sensorial dimension that refers to bodily ecstasy, sensual intensity, and affective exuberance. But *schwärmen* can also mean to rave about something, to fantasize, to worship, to be enthused about something or someone. It describes an affective state of ecstasy, of being beyond oneself, forgetting one's own individuality, merging with another. The latter is inspirational for my reading of Jünger's literary fantasy about drone swarms, insects, and communities as multitudes.

Insects and Communities

Biologists' observations of swarm formations are often mediated in social-political discourses. For example, entomologist Karl Ritter von Frisch's *An Account of the Life of the Honey Bee* (*Aus dem Leben der Bienen*, 1927) analyzed bees' social interactions, such as the famous "waggle dance," which is used to convey information about distant food sources: "Bees, similar to other social insects such as ants, were hardworking, well-organized, and effective. It was as if everything was in tune in this collective of individual insects, which showed highly concentrated social behavior."[25] According to media archaeologist Jussi Parikka, bees' behavior has offered a social and political trajectory to generate models of human communities. Even in classical political theory, bees were often contrasted against Western political and cultural models: "Aristotle's political animal is contrasted to insects; Hobbes saw the insects as incapable of sovereignty; Marx for his part explained that the human being is superior in its capacity to extrapolate, abstract and idealize in his labouring activity."[26] In another famous example, Bernard Mandeville took the beehive as a liberal model for the constitutional monarchy. His *Fable of the Bee: Or, Private Vices, Publick Benefits* (1714) describes a bee community that lives in prosperity; but once the community decides to live by honesty and virtue, its economy collapses, suggesting that selfishness and private vice can be productive of public and economic benefit.

Modernist visions of insect communities are frequently based on hierar-
chical and authoritarian social order. For example, in Maurice Maeterlinck's
Life of a Bee (1901), the individual bee is overruled by the collective beehive.
This view of the bee community also resonates in Waldemar Bonsels's book
The Adventures of Maya, the Story of an Industrious Bee (*Die Biene Maja und ihre
Abenteuer*, 1912), later adapted into a film and a cartoon.[27] The individualis-
tic spirit of the little bee Maya comes into conflict with the strict rules and
hierarchical (even totalitarian) power relations of the hive. These modernist
insect community visions are characterized by a "politics of rationalization
and birth of a certain 'Fordist order.'"[28] Inspired by capitalism's doctrine of
efficiency, these communities are defined by a central authority that creates
hierarchical, top-down social structures.

Parikka argues that this connection between bee communities and hierarchical community models took a new turn with the emergence of cybernetics, a heterogeneous field of systems research that encompasses biology,
sociology, engineering, communication studies, and many other dis-
ciplines.[29] The rise of cybernetics is often connected to mathematician
and engineer Norbert Wiener's work on self-regulated systems.[30] Parikka
shows how important animal behavior was for Wiener's cybernetic theory,
which often compared the sensing of animals to the sensing of machines:
Parikka discusses the experiments in which Wiener used moths to investigate light-sensitive behavior.[31] The French scientist, physiologist, and photographer Étienne-Jules Marey had already experimented with the connection
between insects and self-steered technology as early as 1873. His artificial
insect machine, *La Machine Animale*, investigated flight movements, using
the insect as a sensorial and physiological case study for the mechanical
apparatus.[32]

During the twentieth century, sensorial explorations of insect behavior
continued to accrue societal and biopolitical dimensions: "Insect bodies
contract and transduce more abstract social and political concerns; insects
mediate. . . . Insects could offer, in addition to the lessons in rationalized
management of a Taylorian kind, lessons in agencies without a center, phenomena that during recent years have been raised as the new modes of radical politics through concepts such as multitude, swarms, and smart mobs."[33]
Insects can organize alternatives to Fordist models, and to hierarchical and
sociocentric social configurations; they can offer social agency without a
center. What alternative models can we learn from insect swarms, and even

from drone swarm imaginaries? In what follows, I analyze a work of fiction that provides an example of such an alternative model—interestingly, one written during the same period as the Macy Conferences, which explored cybernetics as a transdisciplinary field of knowledge production.

Ernst Jünger's Drone Swarm as Multitude

In cultural and aesthetic research on drone imaginaries, Jünger's science fiction story *The Glass Bees* (*Die gläsernen Bienen*, 1957) has become widespread currency. Scholars have discussed the German author's hallucinatory fantasy about swarms of microdrones in terms of the aesthetics of terror, violence, and warfare.[34] A soldier and writer, Jünger (1895–1998) lived a long and controversial life. His conservative, antiliberal attitudes, his aestheticized writings about war and violence, and his fascination with the National Socialist movement in Germany still polarize scholars today.[35] Jünger had a deep interest in technology, especially in automation, communication, and visual media. In 1949's *Heliopolis: Retrospective of a City* (*Heliopolis: Rückblick auf eine Stadt*), a dystopian novel about a future metropolis, he describes the phonophor, a mobile communication device that might be interpreted as a forerunner of the cellphone; in the postapocalyptic *Eumeswil* (1977), the Luminar suggests an early vision of the Internet. Jünger was strongly interested in emerging discussions about cybernetics and was likely aware of the Macy Conferences.[36] *The Glass Bees* connects his interests in robotics, emergent systems, and insects. As is well known, Jünger studied zoology, marine biology, and botany, and he was a passionate entomologist. His interest in insects inspired his poetic writings, and his imaginary insect worlds often coincided with the worlds of war. In *Heliopolis*, several passages in the chapter "The Apiary" ("Das Apiarium") focus on the military or battlefield potential of bee swarms, describing "the murder of the queen, the duel between queens, or the account of the drone battle."[37] His famous World War I memoir *Storm of Steel* (*In Stahlgewittern*), first published in 1920, teems with insect analogies: battlefields like bee swarms, soldiers like ants, artillery fire like insect noises. This "insectization" of war highlights Jünger's nonempathic view of violence. Hans Blumenberg grasps this by describing Jünger as a "moon man," someone who sees human life at a distance.[38] But despite the clear intersection between military and insect organization in Jünger's prose, I wish to read *The Glass*

Bees beyond a military interpretative framework. Jünger's microdrone swarms can tell us something about societal organization other than the vision of a dehumanized insect state. Instead, I will argue that the artificial drone swarm and its sensorium in *The Glass Bees* suggests a community vision of the multitude. Given the political controversies about Jünger, my endeavor may appear highly unorthodox. But in addition to exploring the drone sensorium, my analysis also seeks to provide a counterperspective to the scholarship that fetishizes Jünger as an author of aestheticized terror.

The Glass Bees describes robotic bee swarms that are barely distinguishable from natural bees. They are an early (perhaps even the first) literary imagining of microdrones. These drone bees are designed by the automata maker and big-tech industrialist Zapparoni. The narrator, Captain Richard, an unemployed cavalry officer, has a job interview with Zapparoni and visits the company compound. The Zapparoni company houses the smartest minds and inventors, whose work conditions are rather unusual for this historical period: very high salaries, a lot of individual responsibility, and constant surveillance. In some ways Zapparoni foreshadows the "alternative" business philosophy of Apple today, or perhaps the eponymous company in Dave Eggers's 2013 novel *The Circle*, a hip young mega-enterprise with total surveillance. Zapparoni specializes in minirobots and automata, and there are literally swarms of them: "Swarms of selectors could not only detect the faintest smell of smoke but could also extinguish a fire at an early stage; others repaired defective wiring, and still others fed upon filth and became indispensable in all jobs where cleanliness was essential."[39] Even Zapparoni himself has similarities with a robot. When Richard first meets him, he is struck by an uncanny sensation as he looks into his "synthetic blue" eyes.[40]

In the course of the novel, Richard becomes acquainted with Zapparoni's latest invention: glass bees, insectoid microdrones. While waiting in a picturesque garden at the Zapparoni compound, he observes swarms of bees humming and buzzing amid the flowers. He soon discovers they are artificial and robotic: "I distinguished diverse models—almost colonies—of automatons which combed the surrounding fields and shrubs. Creatures of especially strong structure bore a whole set of proboscises which they dipped into umbels and flower clusters. Others were equipped with tentacles that closed around the tufts of the blossoms like delicate pincers, squeezing out the nectar. Still others remained a puzzle to me."[41] Certainly, these robotic bees qualify as drones. According to Richard's observations, they are a little bigger than natural bees, and they are transparent. They flutter like bees, they can hover (flying in "wide figures of eight"[42]), and they seem to be steered remotely (although we never find out who is steering them). They are also equipped with visual sensors that enable them to find flowers and maneuver through the garden without collisions. Richard speculates that they may have the ability to transmit signals ("cell transmitting orders"[43]) and thus to communicate with each other by means of some type of electronic network. Their artificial wings, which make a constant sound ("hum of the bees"[44]), are used not only for orientation within the swarm but also to trigger pollination. Thus the glass bees feature a nonhuman drone sensorium that combines zoological and machinic ways of sensing

Richard finds this sensorium difficult to understand. He is perplexed by these automatic bees, which confound his ability to differentiate between fact and fiction. Everything might be a hallucination. But the microdrone swarms in *The Glass Bees* do more than push the narrator to the limits of his cognitive capacity to understand the world. They also suggest a puzzling form of community based on zoological units, cybernetic visions of technology, and decentralized principles of power. For Jünger, the bee is a political being. He had already shown this in *Heliopolis*, whose narrator mentions that the bee used to be a state animal and to serve emperors—hence its appearance as a heraldic emblem.[45] However, the bee as a metaphor for absolutist sovereignty—an age-old trope, as discussed earlier—no longer quite makes sense in *The Glass Bees*.[46] So what kind of community do these swarms of glass bees suggest?

At first, Jünger's automated, translucent glass bees seem to exemplify modernity's automatization of human lives. Richard laments the general mechanization of humanity. A man of "yesterday," he dwells in nostalgia for the time when war was fought on horseback and the metric system had not infiltrated every aspect of life.[47] We also find such observations about the mechanization of modern life in Jünger's aesthetic and political theory. His 1932 essay "The Worker ("Der Arbeiter") proposes a new, dividualized type of worker as part of an authoritarian model of community. This worker is characterized by nonempathy: like an insect, the worker takes a cold, distanced view of the world that Helmut Lethen calls a "gaze of heights."[48] The worker's technology is photography, whose cold and affectless gaze trains workers for emotional numbness and prepares them for war.[49] We might see

The Glass Bees as a sci-fi vision of this new type of worker. The robotic bee swarm can be read as a model of this dehumanized community, organized by hierarchies and on military principles. However, *The Worker* is separated from *The Glass Bees* by more than twenty years, and Jünger's writing and ideology changed after World War II.

At first Richard understands the drone bee swarm as a community centralized around a single powerful principle. The bees are mere mechanical units in a larger system: "Although scores of units were involved, the whole process was conducted with perfect precision; no doubt, some central control or principle regulated it."[50] But he never finds that central principle. Indeed, he becomes increasingly bewildered as he observes the swarming of the bees: "The way they radiated from the hives in clusters, threw themselves like a glittering veil over the display of bright flowers, then darted back, stopped short, hovered in a compact swarm—from which, by inaudible calls and invisible signs, the gatherers, one by one, were swiftly summoned to deliver their harvest—all this was a spectacle which both enthralled and mesmerized."[51] He seeks the central voice of the swarm, but it is controlled by invisible, inaudible commands that have no origin. Richard wants Zapparoni to be an authoritarian figure like the hacker in "Hated in the Nation," the single power behind the swarm. But he is not. Instead, Richard discovers that the bee community functions as a *multitude*—a term that needs a short theoretical explanation.

In French critical theory, insects have attained status as a metaphor to illustrate an organizational basis for practicing community. Gilles Deleuze and Félix Guattari see ants as able to form a *rhizome*, a thought figure that allows for multiple entry and exit points, openness, and nonhierarchical connections: "You can never get rid of ants because they form an animal rhizome that can rebound time and again after most of it has been destroyed."[52] Michael Hardt and Antonio Negri follow this rhizomatic order of ants; for them, the ant is a model of collective intelligence that is embedded in the *multitude*.[53] This concept of multitude has its roots in modern political thought and can be found in the work of Niccolò Machiavelli, Thomas Hobbes, and Baruch Spinoza.[54] Different thinkers understand the concept in very different ways, and it involves many tensions and contradictions: "At times the multitude is used as a synonym for what we might call the 'masses' or even the 'people,' while at other times the multitude is given a very specific political charge as the constitutive force of social and political life itself."[55] Like the swarm, the multitude is inherently paradoxical: it consists of singularities, but those singularities share something in common. For me, the guiding idea here is that multitudes are constituted by social energy and affective relations. As Spinoza himself highlighted, affective energy can be conducive to the formation, envisioning, and construction of multitudes.[56]

Hardt and Negri describe the multitude as an unmediated, nonhierarchical, and nonbinary collective social configuration. The multitude is neither the individual nor the group; it is a whole that consists of singularities. The multitude has no social contract and is not constituted by any deliberative consensus. It suggests the idea of a common, but this common is constituted in a fluid, self-emergent network, relating affects, themes, and experiences to each other. Thacker summarizes some of its main characteristics: "The multitude is a whole of singularities. Unlike the people, the multitude is not a unity, but as opposed to the masses and plebs, we can see it as something organized. In fact, it is an agent of self-organization."[57] Consequently, the concept of the multitude can give us a vision of community that comes close to Jean-Luc Nancy's vision (discussed in chapter 1): a relational being-together, without the sovereign power that subsumes singularities.

To return to *The Glass Bees*, the question remains: What community model do the microdrones and their sensorium suggest? Richard desperately tries and fails to locate the central power that steers the swarm. He concludes that the artificial bees are centerless. Their hives do not really look like apiaries, but more like stations in an "automatic telephone exchange."[58] The hives' "entrances functioned rather like the apertures in a slot machine or the holes in a switchboard."[59] The bees communicate via hivelike rhizomatic networks. The communication happens everywhere and nowhere, in an invisible cloud comprising "scores of units"[60] that are plugged together and networked in a cluster of contact points. Richard finally decides, "One could hardly assume the existence of a central control panel: such a device would not be in the Zapparoni style because for him the quality of an automaton depended on its independent action."[61] Thus the swarm sensorium of the glass bees embodies a nonhuman zoological form of sensing that is based on decentralized networking and communication.

Richard's struggle to understand this centerless swarm sensorium is particularly evident when he describes its sonic dimension. When an artificial bee swarm occurs, there is a constant droning and buzzing in the air.

Reading Jünger's text is like diving into a sound chamber of white noise. The Zapparoni company garden is pervaded by a continuous humming ("the air was filled with humming"[62]), whistling ("high continuous whistle"[63]), and swarm sounds. This combines with the "steady roar from the plant, the rumble from the parking lot," and the "murmur from the treetops."[64] When Richard first experiences the sensorium of the drone swarm, he falls into a state of ecstasy and "enthusiasm."[65] After a while, however, the undifferentiated cloud of noise leaves him utterly confused. The bees' white noise is the sensorial matrix of a centerless multitude. There is no leader principle; there is no queen. The robotic beehives suggest flat communities without hierarchies, class structures, or power differences. Not only does the swarm seem to communicate and format itself without a center, but it also eliminates binary gender: "The whole establishment radiated a flawless but entirely unerotic perfection. There were no eggs or cradles for the pupae, and neither drones nor a queen."[66] Unlike Richard, Zapparoni—a rather artificial, sexless, and ageless figure—has long understood the antibinary logic of the multitude. He is himself beyond binaries. Indeed, the whole company seems to rely on a swarmlike, multitude form of organization. Zapparoni tells Richard, "Do call me simply by my name as all the workers in our plant do. He did not say my plant or my workers."[67] This decentralization is mirrored in the architecture of the plant, which is built on the ruins of a monastery and has a complex, rhizomatic tunnel system, with numerous exits and entrances that allow the plant to be entered from many directions. Like the glass bees, the swarms of small humanoid robots with which Zapparoni stages great filmic spectacles also seem to have a certain degree of autonomy and free will. Richard says, "It was a luxury puppet show without puppeteers and wires, not only a new play but of a new genre."[68] Yes, they are playing a new genre: they are practicing multitude.

Richard is not ready for all this. He resists the multitude by having a nervous breakdown. He (perhaps understandably) finally loses his composure when he discovers another swarm, namely, masses of severed ears lying on the muddy ground: "Now I started scouring the water hole methodically and with increasing horror: it was dotted with ears. I distinguished large ears and small, well-shaped and ugly, and all had been severed with neat precision."[69] The field of ears—a hallucinatory, psychedelic scenario, perhaps inspired by Jünger's experiments with LSD[70]—push Richard over the edge, and he loses "his nerve."[71] But the swarm of ears is only the logical consequence of the idea of the multitude, which dispenses with the body as an anthropomorphic, human-individual-centered entity. The ears suggest a body that is scattered and departmentalized. In *The Glass Bees*, the body becomes, as Devin Fore puts it, a construction site and a decentered form of assemblage that is connectable to nonhuman forms, such as insects and automata.[72] The severed ears symbolize a body assemblage that is connectable to nonhuman (technological and zoological) configurations of communities.

This ontological deconstruction sends the nostalgic and old-fashioned Richard into a panic attack, and like the soldier he once was, he starts a quarrel with a bee he calls Smoky Gray. This walnut-sized bee, which is a little bigger than the others, has been circling and hovering around him the whole time he has been in the garden. Richard feels threatened by Smoky Gray and hits it with a golf club. The robot bee explodes and squirts a liquid onto him: "I hit the Smoky Gray with the flat end of the iron and smashed it. I saw a coil of wire spring out of its belly. Several sparks followed, as if a toy frog were exploding, and from the iron golf club rose a rust-brown cloud. Again I heard a voice: 'Close your eyes.' A splash hit me, burning a hole in the sleeve of my coat."[73]

This scene has been interpreted as a violent act that mimics Jünger's traumatic memories of World War I's mechanized battlefields.[74] Such a reading certainly makes sense given that the liquid seems to be acidic and dangerous. However, if we put this scene into the context of the swarm's multitude community model, the splash can be read as a form of fertilization to shape a new human that will overcome the subject–nature divide: a human that can cope with hive minds, swarm collectives, and autonomous automata. Christine Kanz has shown that Jünger's prose frequently refers to male birthing fantasies, and Smoky Gray may be another example of this.[75] The product of the birthing process is not a super-soldier, but a type of human that can become a construction site adaptable to the multitude, a cyborg that can be assembled with the parts of a machine. Of course, Jünger's cyborg vision can be read as a problematic echo of the new worker type outlined in his profascist anthropology. But the newborn Richard's body is not a heroic worker machine, it is a disassembled construction site that unites the drone bee robot with the human. The human has been decentered; ontologically, it is no longer positioned above technological and zoological modes of being.

The Glass Bees suggest that the human body should be compatible with the body of the swarm as multitude and should become a part of it. There is no need for a central intelligence; the swarm does not execute destructive power. As Richard himself admits, he is not even really sure that the bees are dangerous: "I could not even contend that the Smoky Gray had an evil design on me. I had lost my nerve, as the saying goes."[76] In this light, the community vision projected via the sensorium of Jünger's drones does not compute with a totalitarian society of robotic killer bees. Rather, the swarm embodies a decentralized multitude, and the zoological sensorium of the drone swarm suggests that technology might shape visions of communities that are not entirely controlled by human authority. This does not place Jünger beyond political controversy, but his prose operates as a seismograph of assemblages of human agents, animals, and technology, and their ramifications for social imaginaries.

The Politics of Swarming and Drone Swarms

So far I have discussed artistic representations that show how drone swarms sense, and how their sensoria suggest imaginaries of communities: Netflix versus Jünger; swarm as violent dehumanized community versus swarm as experimental multitude. In these artworks, the swarm and its communities have remained metaphors that invite reflections on technology, zoology, and social figurations. But drone swarms can also configure real-life communities. In this section of the chapter, I concentrate on domestic drones that shape communities through deployment by protest groups and political activists. Political activists' use of drones as a medium of protest is no longer a novelty. Think of the drone footage of Black Lives Matter murals in the five boroughs of New York City. Or consider the Mexican artist collective that initiated the Rexiste movement to protest the missing Ayotzinapa students and perform actions against former Mexican president Enrique Peña Nieto. Rexiste uses the protest tool Droncita (Little Drone), an airborne robotic graffiti drone programmed to spray red paint on public images of Peña Nieto. Droncita can efface images of the president's head, and its actions are distributed across social networks and websites.[77] Austin Choi-Fitzpatrick has shown how social movements use civilian ("good") drones to document human rights abuses, crowd sizes at demonstrations, and climate change research.[78]

This section of the chapter adds an aesthetic aspect to this discussion by focusing on activist drone art and the drone swarm as a sensorium.

William Connolly has coined the expression *politics of swarming* as part of his theorizing about the planetary, humanism, and activist movements.[79] He discusses the urgency and necessity of political action in light of the impact of climate change and global capitalism on the human community. Connolly pursues the idea of organizing nonviolent protest against states, institutions, and corporations to push for radical reforms to infrastructures of consumption and ecodestructive processes. For him, the metaphor of the swarm describes how activism and resistance are organized in today's digital and networked society. Connolly refers to Thomas D. Seeley's *Bee Democracy* (2010) to make his point: "Seeley construes the beehive on the model of the human brain, another decision-making assemblage without a central coordinator."[80] In line with my previous discussion of swarms, Connolly sees the swarm as a collective with centerless intelligence. He considers the swarm to be an analogy for communities, since current social protest movements also organize themselves in this way. Activists no longer organize themselves around one platform, central voice, medium, or location. Instead, swarm units are involved in simultaneous, at times contradictory, network activities, coalescing people from different regions and social positions: "The politics of swarming, then, is composed of multiple constituencies, regions, levels, processes of communication, and modes of action, each carrying some potential to augment and intensify the others with which it becomes associated."[81] With these words Connolly describes the dynamic of smart mobs, groups that are empowered by digital networks and can mobilize fast via social media. The collective goal of the politics of swarming, according to Connolly, is to set the stage for a "cross-regional strike"[82]—most importantly for him, a climate strike that would resemble workers' strikes but whose goal is to stop consumption and cultivate an "ethos of positive material austerity."[83] The swarm embodies an emergent volatile energy, and although decentered, it can birth a common action or intervention.

Drones can be technological players in this politics of swarming. A great example is the drone disruptions in 2018 at Gatwick, one of the UK's busiest airports, where 140,000 passengers on a thousand flights were affected. It is still unclear how many drones were actually in the air, but they were reportedly seen and interestingly even imagined as swarms. It all began like the flocking scene in Hitchcock's *The Birds*. At first only two drones were spotted around the airport. After a while, more and more reports of drones trickled in. Eventually 170 people claimed to have seen them; 115 of these reports were endorsed as credible by the police.[84] Gatwick was closed for days. A Drone Dome was even ordered, a military device to snatch drones out of the air.[85] The pilots behind these disturbances were never caught (two people were wrongly arrested). At Gatwick the drones did not fly in intricate swarm formations, they popped up in groups. Nevertheless, they had swarm elements: their occurrence remained unpredictable, and no central controlling pilot was ever identified. Their main effect was to disrupt air traffic by their sheer appearance.

Inspired by the impact of this incident, the environmental activist group Extinction Rebellion planned to fly a swarm of drones in the Heathrow area on specific days in June 2019 to protest against the building of a third runway. They appeared confident that they could close Europe's busiest airport for a minimum of sixteen days, disrupting the travel plans of at least 3.5 million passengers. They stated, "There is nothing violent about flying drones when there are no flights in the air as it is perfectly safe. We are there first. The responsibility is with the airport authority to not initiate flights."[86] Extinction Rebellion and its splinter group, Pause Heathrow, ultimately did not go through with the action. The reasons may have been the threat of a life sentence or the riskiness of the action itself.

For my purposes in this chapter, this protest action is obviously an interesting case. Extinction Rebellion wanted to use drone swarms, and at the same time it is a community that sees itself as organized like a swarm: "We organise in small, autonomous groups distributed around the world. These groups are connected in a complex web that is constantly evolving as we grow and learn. We are working to build a movement that is participatory, decentralised, and inclusive."[87] For better or worse, the drone has become part of a social movement, a medium of the politics of swarming; in Choi-Fitzpatrick's words, drones work as a "technology of resistance."[88]

Another case that illustrates the politics of swarming with drones is the protest movement against the Dakota Access Pipeline. In these protests, drones were involved not as a threat but mainly to document the protests and as an artistic medium. The movement began in 2016 to protest plans for a pipeline that would deliver oil from western North Dakota to southern Illinois. This pipeline would cross Lake Oahe and the Indigenous territory ("reservation") of Standing Rock, constituting a threat to drinking water and cultural sites. Many different protest groups gathered at the Sacred Stone Camp, a common ground for Indigenous people and political activists. The camp and its inhabitants were subjected to constant state and police surveillance during the protests. In addition, the police sprayed demonstrators with ice-cold water in freezing temperatures, and they used tear gas and dogs. In his study of the protests, Nick Estes compares the situation to a military occupation: the camp was completely encircled by law enforcement and surveillance.[89] State police and private security firms surveilled the protesters not only from the ground but also from the air, using domestic drones. J. D. Schnepf describes this surveillance as "atmospheric policing"[90] by which the state authorities gained colonial sovereignty over the airspace. The Federal Aviation Administration even passed temporary legislation to ban all private drone flights from the area. The protesters countered the police violence with a swarm of their own drones, filming and documenting the government's actions. Many different resistance groups, including ReZpect Our Water, the International Indigenous Youth Council, and Digital Smoke Signals, as well as many Twitter fora (for example, #NoDAPL tweets), documented and reported on the resistance and its use of drones.

The artist Frédérick a. Belzile used a drone to create an aesthetic work about the events at Sacred Stone: a three-minute film entitled *Eyes in the Sky* (2017). In black and white, the film shows the camp from the viewpoint of a hovering drone. The images display an abstract quality, more reminiscent of a drawing than a film. Some black areas are filled with a snowlike drizzle, recalling the snowy static on an old television screen. The contours are fuzzy, and the gaze from the drone is blurry. This visual aesthetic is remarkable when we consider that the drones used at Standing Rock were probably supposed to deliver high-definition footage to document the ongoing protests and police violence. The drone in *Eyes in the Sky* is not one that views according to extreme mastery or shoots its scopic regime vertically into the ground. Rather, the drone seems lost, insecure, flying backward at times. This drone sensorium also sets up a counterpoint to the police surveillance by taking sides with the protesters, telling and showing their narrative.

The narrator enhances this feeling of disorientation, as the voice-over relates that the drone got lost, went rogue, and flew off course. The footage and narration are taken from live feeds and posts by the protesters; the voice of the narrator belongs to an Indigenous drone pilot. He begins the narration with these words: "I'm taking a moment of silence because the Indigenous drone flew away. And I probably never ever see this drone again. Because it just flew away. The screen went blank and I pushed Home, to make it come home. And the drone just flew away."[91] He speculates that it may have been hacked or captured, or even shot down by law enforcement. His tone is sad, complementing the melancholy of the images in the drone film. The film mourns the violence that was done to the people and will be done to the land. But it does this in alliance with the drone. For the narrator, the drone becomes an emotional and empathic interlocutor: he cared for the drone and loved it, and now he mourns it like a lost relative or friend. The drone in this film thus embodies a different type of "camera consciousness,"[92] one where the drone is no longer the perpetrator, but rather is an affective technology that can trace the protests. Its going rogue symbolizes the loss, decimation, and disappearance of Indigenous territory.

Both *Eyes in the Sky* and the various protest groups at Standing Rock used domestic drones as an act of atmospheric sousveillance (see chapter 3), appropriating tools and technologies for their resistance struggle.[93] Although the protest drones did not fly in sync, we can still see them as constituting a web—a swarmlike shield against the state's constant atmospheric policing. This is particularly the case insofar as the images from the protest drones were immediately uploaded and channeled through numerous online networks, such as Twitter, Facebook, Reddit, and other public media. The drone swarm thus embodies the multitude of the protest community. Just as the drones are not steered by one center, so too the various protest groups cannot be seen as a hierarchical organized community, but rather as assemblages of units connected via networks, digital and social media, and drones. Here the idea of the multitude gains a more concrete political dimension than it has in the aesthetic drone imaginaries of "Hated in the Nation" or *The Glass Bees*.

Schnepf has voiced skepticism regarding the creative reappropriation of drone technology as a tool of swarm resistance, on the grounds that drones are always already imbricated with global capitalism, consumerism, and militarism.[94] But while I agree that it is always necessary to remain aware of where drones come from, who owns them, and how they are produced, it is not productive to always keep clinging to this genealogical fallacy of the drone. Although the pipeline was eventually built, the use of drones certainly intensified the protesters' political power. The drone's capacity to move and communicate independently enabled the protest to be communicated in large networks: "The drone's motility, its autonomous vertical and lateral movement differentiates it from the mobile camera as it generates distributed modes of vision across a number of bodies, devices, and platforms, not merely as spectacle, but as a new mode of relational experience."[95] In other words, this relational experience of the drone was articulated by the swarming features of the drone and its networks. The protesters' drone swarms pursued the politics of swarming, a decentered mode of resistance consisting of an assemblage of multiple agents, fora, and technologies. Of course, we have to be careful not to idealize these Indigenous community drones, since drones also surveil networks in ways that can be used against protesters. The drones at Standing Rock nevertheless invite us to follow the politics of the swarm by enhancing participation, connection, and networking beyond the paradigm of vertical violence.

Swarm Communities

Drone swarms feature the "zoosensoriality"[96] of technology, meaning that they have adapted animal modes of sensing. The drone as swarm thus adds another nonhuman element to the drone assemblage. This chapter shows that the drone tends toward the animal world. The military is keen on naming drones after animals, as names such as Killer-bee, Black Hornet, Wasp, and Hawk illustrate.[97] Drones adhere to the sensorium of swarming, executing insect vision, networked perception, and centerless communication. Drone swarms suggest a form of rhizomatic sensing. They offer a sensorial and aesthetic medium through which to imagine nonhuman-centered communities, such as multitudes and decentralized collectives. As my discussion of "Hated in the Nation" shows, sometimes the zoosensoriality of the drone does lead to a militarized "technobestiary"[98] that annihilates and dehumanizes communities—but it does not have to do so. Jünger's drone zoo suggests a community model of the multitude as a creative posthuman figure of thought. Although Jünger is often connected to the aestheticization of war and violence, *The Glass Bees* delivers a speculative matrix through which we can reflect on drone swarm communities in a posthuman framework. But it is not only in the filmic and literary world that drones can organize swarmlike collectives. Protesters and social activists use and reappropriate drone swarms for their own goals. As I have shown, drones and their ability to swarm can become actors in social movements. Whether these drones are "good" drones depends on one's political point of view. The Gatwick incident and planned Heathrow protests demonstrate that for some, the use of drones is designated as terror, while for others, drones are a necessary and ethical medium of resistance in light of the climate crisis. Be that as it may, these works and movements show that drone swarms are not always tied to network-centric warfare. Their nonhuman machinic and zoological sensorium can inspire creative minds to experiment with community models that dare to displace the centrality of human control.

In a somewhat eerie way, drones seem to have been waiting for the COVID-19 pandemic as their moment to shine. A *New York Times* article dated May 23, 2020, expresses it well: "The Drones Were Ready for This Moment: While humans hunker below, aerial robots are taking over as cops, street cleaners, medical assistants, delivery workers and soon—friends?"[1] It seems that COVID-19 has empowered drones to shed their homicidal stigma and reveal themselves as aspirational and humanitarian saviors. In Wuhan, drones delivered medical supplies, enabled contactless pick-ups, and even provided light for hospital construction.[2] In the US and Europe too, drones have been fighting the pandemic: as aerial virus blasters, spraying disinfectant to sanitize the streets; monitoring social distancing; sensing people's temperatures and heart rates; detecting coughs and sneezes in human crowds.[3] For example, the drone monitoring of citizens was attempted in Westport, Connecticut, in a cooperation between the police department and the drone company Draganfly.[4] The police department implemented a program called Flatten the Curve, using drone technology to help "the community practice safe social distancing, while identifying possible coronavirus and other life-threatening symptoms."[5] The program was later scrapped when the American Civil Liberties Union voiced its outrage and demonstrated against drone surveillance and the intrusion of privacy.[6] Drones are just one tool in an array of digital technologies to cope with the ongoing health crisis. Tracking apps, digital fences to enforce quarantine, and the use of artificial intelligence (AI) to predict the spread of the pandemic are also examples of technologies intended to contain the virus, and some of these have indeed intruded on people's civil liberties and privacy.[7] The drone has boomed in the time of coronavirus, and its new role as a COVID-19 buster has upgraded its sensorium. Pandemic drones are equipped with thermal, visual, and datafied sensing technology; as I will show, they have even learned to shout at people.

My aim in this chapter is to investigate the expanded sensorium of the pandemic drone, and to discuss how it is shaping imaginaries of us as humans, our relations to each other, and our communities. I focus on pandemic drones used by governments to deal with the crisis, as well as on drones as aesthetic and artistic media to record, represent , and interpret the pandemic. Is the pandemic drone a technology that will ultimately lead (or has already led) to a police-controlled surveillance state? Does the COVID-19 drone present a domestic façade behind which it militarizes our communities? Or can the pandemic drone, and particularly its medialization in artworks, envision new forms of companionship between humans and machines, machinic empathy, and intersubjective machinic communities? While I am more interested in the nonmilitary aspects of civilian drones, the first part of this chapter discusses pandemic drones and their sensoria as phenomena of what Michael Richardson calls "everyday militarism."[8] This discussion establishes a point of departure from which I then venture into aesthetic pandemic drone sensoria and imaginaries, of which the military matrix is no longer the main signifier. Most of my artistic examples here are taken from a B-grade aesthetic realm: low-budget films, amateur drone shoots, and public entertainment. As I mentioned in the book's introduction, for me, "B-grade" does not imply a qualitative aesthetic judgment that situates such artworks as inferior to high culture. Quite the opposite: it is precisely their closeness to the aesthetics of "trash" and dilettantism that provides me with the space to rethink the drone. Many of these amateur drone films experiment with new relationships between the human, the nonhuman (virus), and the drone.

In this context, I have been guided by Siegfried Kracauer's aesthetic theory. Kracauer's interest in the surface phenomena of mass culture as a dialectical, epistemological palimpsest is productive for my readings of the viral sensorium of the drone. At the very beginning of his essay "The Mass Ornament," ("Das Ornament der Masse," 1927) he makes clear that he sees aesthetic "surface level expressions"[9] as keys to knowledge about society. Likewise, I see amateur art with drones as a prism for knowledge about communities and the role of technology during pandemic times. In this chapter, I investigate the 2015 independent sci-fi movie *Rotor DR1*,[10] in which drones become companions to humans in a world devastated by a deadly pandemic. This film and its fantasy of drone companionship lays the groundwork for my subsequent discussion of drones as "empathy machines." Finally, I explore the dialectical powers of mass drone shows, in which drones write messages and images in the sky. In all these aesthetic drone sensoria, the viral sensing modes of the drone are tested for their potential constructiveness regarding the larger implications of the global COVID-19 political and health crisis for our relationship with machines, and for our communities.

Shouting Drones

One of the most peculiar tasks assigned to the pandemic drone is that of shouting at people. Governments are sending domestic drones into the air to enforce social distancing rules by yelling, screaming, or simply talking to citizens. A video that went viral on Twitter shows an elderly woman in a village in Inner Mongolia: a drone is screaming at her for not wearing a face mask. In a video on the website *Global News*, we see this woman staring at the drone and then running away when it starts to shout: "'Yes auntie, this drone is speaking to you,' the voice says, according to translation in a report by China's state-owned Global Times. . . . 'You shouldn't walk about without wearing a mask. You'd better go home, and don't forget to wash your hands.'"[11] In another example, an invisible police officer in the northern Chinese city of Shuyang uses a traffic drone to call out to maskless pedestrians: "Hey handsome guy, speaking on your mobile—where is your mask? Wear your mask!"[12] The tone of the shouting drone is a little less polite in Italy, where the mayor of Messina recorded his own voice onto a drone and screamed at breakers of quarantine: "Where the f*** are you going? Go home."[13]

Even these few examples of drone usage during the pandemic show that the drone has expanded its sensorium: in addition to its buzzing or humming sounds, it can now also shout. Of course, drones with loudspeakers were not new inventions in 2020, but they became more popular that year. Drones have gained the sensorial feature of a disembodied voice that seems to come from nowhere—a voice that can scare people. In all the videos I have examined, the people being shouted at by a drone express fear through their body language. Disembodied voices are deeply uncanny. The French critic Michel Chion has written about the strangeness of a voice with no origin. He calls this "acousmatic," a faceless sound that is heard

Figure 7.1
Photo illustration by Slate. Photo by Miguel Medina. © Miguel Medina.

without its cause or source being seen.[14] Chion refers to the panoptic power of the acousmatic voice: "The *acousmêtre* is all-seeing, its word is like the word of God: 'No creature can hide from it.'"[15] To illustrate the uncanniness of disembodied voices, Chion refers to the supercomputer HAL in Stanley Kubrick's *2001: A Space Odyssey* (1968), and to Fritz Lang's silent film *The Testament of Dr. Mabuse* (1933).[16] We might also think of Aldous Huxley's eerie novel *Brave New World* (1932), in which anonymous commands are constantly whispered through sound boxes. The acousmatic power of pandemic drones is visible in the terrified faces of their targets.

Not only are the shouting drones creepy but their effectiveness is highly questionable. In a crisis that is already alienating people from one another, a disembodied voice does not seem a good solution. Would the public not be more inclined to trust a human being? For me, the shouting drones are symptomatic of the societal tendency to resort to technofix fantasies during a crisis. Indeed, technological utopianism—the idea that the latest technology will put an end to a current state of insecurity, catastrophe, and uncontrollability—is by no means confined to literature and art.[17] There is a tendency to believe that technology will solve societal problems—for example, if we invest in care robots in hospitals rather than hiring more staff. Sometimes, of course, the technological solutions are the most efficient, but they can also turn out to have drawbacks in the longer term.[18] The technofix is often treated as a substitute for long-term political action, and it overlooks the systemic dimensions of the problem in question. For example, shouting drones might be a quick way to enforce social distancing rules, but they do not tackle the structural problems at stake in COVID-19: the high death rates among the poor, the inequalities embedded in health care systems, or the connection between climate change and humans' increasing vulnerability to pandemics. Thus I am skeptical of the "help" that these shouting, spraying, and tracking drones can really provide during the pandemic, compared with the dangers they pose regarding privacy and surveillance.

The shouting drone also reveals its own embedment in militarism. As in drone warfare, the shouting drone removes the drone pilot from danger, as the pilot is at no risk of exposure or contamination. The viral battlefield merges with the everyday and the domestic. The COVID-19 outbreak prompted many governments to formulate and execute emergency measures: stay-at-home orders, social distancing rules, the monitoring and tracking of citizens, and intensified surveillance—orders that would usually apply only in the exceptional case of war. Researchers have described these governmental measures as steps toward a militarized police state.[19] The COVID-19 drone is thus symptomatic of "everyday militarism" insofar as it is used in the domestic sphere but produced by the military industry. Shouting drones are often Mavic 2 models, manufactured by the Chinese company DJI.[20] This company has been accused of spying, but the US armed forces have been loyal customers, buying smaller drones for security purposes even after the spying was uncovered.[21]

The use of domestic drones as militarized "saviors" during epidemics predates 2020. A prominent case is the Ebola outbreak that started in Guinea in 2013.[22] By 2014, President Obama had declared Ebola a security issue, and this converged with drone projects to deliver medicines, measure body temperatures, and monitor sick people remotely. Kristin Bergtora Sandvik notes that the Western response to Ebola and the "deployment of AFRICOM (the US Africa Command), for example, can be viewed as a militarized medical response patterned on the war on terror."[23] Ebola became the target in a kind of remote warfare that demanded the distancing of contacts, interactions, and risks: "The 'war on Ebola' soon proved to be fertile rhetorical ground, both for observers who took the metaphor at face value and for critics who decried the endless use of war metaphors."[24]

Besides this militarization of Ebola, Sandvik further argues that the use of drones during the Ebola crisis only benefited the industrial complex and did not tackle deeper problems. For example, she points to the "leapfrogging" that occurred when the financing of high-tech drones became more important than the building of roads and infrastructure on the ground.[25] The idea of investing in Africa with drones once again constructs the continent as a "site of underdevelopment"[26] for the West's colonialist and imperialist agenda: "The theory is that drones can not only help Africa move beyond insecurity, colonial ills, and humanitarian crises, but can prevent it from languishing in the immature stages of capitalism."[27] Similar criticisms have also revolved around Zipline, a US delivery company that designs and manufactures drones for distribution centers. For example, in Ghana and Rwanda, Zipline specializes in the delivery of blood samples.[28] In 2021, Zipline announced that it would begin delivering COVID-19 vaccines via drone.[29]

In many ways, I agree with all these criticisms of humanitarian projects with drones. However, I also believe that it is possible to improve the world before we save it, and Zipline and other attempts to tackle epidemics can help to ameliorate dire situations. Insofar as "better can be good enough," such initiatives can do some good, even if they do not solve the systemic problems in which the technologies are entrenched. However, this book is not about whether drones are good or bad, or about where the ethical parameters and implications of drone technology lie. Rather, I want to establish what we can learn from pandemic drones and their aesthetic sensoria. Is the only lesson these drones can teach us that they facilitate a surveillance state where the military dictates our way of life, and that all humanitarian drone initiatives have an unacceptable imperialist and colonial agenda? On a pessimistic day, I might say yes: even when they appear to be humanitarian, drones are always complicit with the military and the surveillance industry. But on other days, I am not willing to stop there. I am striving to find out more about the constructive potentials of the drone sensorium, and I believe that pandemic drone sensoria can enable us to reflect on the pandemic's long-term impacts, provide us with new knowledge about relations between humans and technology, and show us the vulnerability of human civilization.

The realm of art and aesthetics offers me this perspective because it so freely experiments with, imagines, and fantasizes about technological scenarios. In what follows, I show that the aesthetic sensorium of the drone plays an important role in dealing with pandemics. Pandemics have inspired many popular disaster movies, including *Contagion* (2011), *Twelve Monkeys* (1995), and *28 Days Later* (2002), to name but a few. Movies where drones play a key role in coping with a pandemic are rare. But I have found one, and it is especially interesting: it was released in 2015, long before COVID-19 defined human life and death on the planet, and it portrays drones beyond good-bad or domestic-military dichotomies.

Drones as Companions during Pandemic Times

Chad Kapper's *Rotor DR1* shows a world destroyed by a pandemic. There are only a few human survivors. Water, electricity, and infrastructure have almost vanished. The only working machines are the autonomous drones that still fly around because no one ever switched them off. Some human survivors, called fringers, scavenge the drones for their valuable parts: arc pellets, which have become the new currency in this postapocalyptic world because they contain electricity, a scarce and much-needed resource. The film follows sixteen-year-old Kitch's search for his father, a biochemist and virologist who is suspected of having botched the vaccine program.

I remember my first viewing of *Rotor DR1* in 2018, three years after its release. I was already working on aesthetic drone imaginaries, but I thought the film lacked the aesthetic complexity that would make it worth exploring for my research. I was put off by its many illogical plot twists. I later learned that the movie had been written using community feedback on Facebook. The facts that the movie's director was the CEO of the drone company Rotor Riot and that the lead actor was his son were also red flags as far as I was concerned. But by 2020, things had changed. The fact that a drone company had made the movie now spoke to my fascination with the dialectics of the drone. Today the movie seems almost prophetic in its thematization of the popularity and apparent benignity of drones during the coronavirus pandemic. But *Rotor DR1* is also intriguing because it does not fetishize the drones as a technofix for a humanitarian crisis. Instead, its drones are sensitive, humanlike companions that cannot solve everything.

Of particular interest is the domestic drone that Kitch finds, the eponymous Rotor DR1. A friendly robot reminiscent of *Star Wars'* beloved R2D2, DR1 becomes a companion for Kitch and his friend Maya as they make their way through a devastated world full of bandits and criminals. DR1 has an extraordinary sensorium, with seemingly endless electric power and a wide array of sensorial features. Besides its many cameras, it can react to motion and has inbuilt AI software (given that the movie dates from 2015, the AI is impressive). It eventually learns to speak, and it seems to have emotions, since it reacts with fear and attachment to its owner. Kitch in turn becomes emotionally attached to DR1 and also develops a close bodily and affective relationship with it. His body and senses merge with the drone in a drone race scene where together they win first place. With his video graphics array (VGA) goggles and his intuitive grasp of DR1's sensorial capacities, Kitch overcomes all obstacles during the race—a symbiosis that recalls the drone race embodiment I discussed in chapter 2.

But it is not only DR1 who develops into a close friend. The other drones in the film also seem capable of companionship. With their highly developed sensorial capacities, these drones were supposed to save the world from the pandemic by delivering vaccines and medical supplies for Skymedics, the company where Kitch's father worked. Indeed, the drones were about to distribute the vaccine when a terrorist group of antivaxxers invaded Skymedics and destroyed it. The terrorist group believed that the pandemic had been visited on humanity by God, and that the strong needed to survive the weak. (As I write this in 2021, this plotline does not seem so farfetched.) In other words, the drone was not a violent technology used by terrorists; instead, the drones themselves fell victim to the terrorists' destructive acts.

This is an interesting shift of perspective for an artwork on drones. In popular films, drones are almost always portrayed as killer machines and surveillance trackers for militarized regimes. This movie is very different. These drones are as vulnerable as humans; they are not fetishized as an infallible technology. Humans and drones need each other. When both the drone machine and the human accept their fragile dependence, they can work with and benefit from each other. The final scene shows DR1 (and the other drones) as an integral part of the rebuilding of the planet's human community. This is not a community that is reminiscent of a police state, but a sci-fi fantasy community where humans and machines live in mutual companionship. The narrator at the end of the movie says, "The world as we know it may have ended but that does not mean that we shouldn't move forward. . . . Because for everyone who comes after us, this will be the only world they ever know."[30]

No doubt this ending is painfully kitschy, but I am willing to embrace it in order to gain a different perspective on drones. The kitsch opens up a fictional fantasy in which drones become part of the human community in a peaceful way. The movie projects an imaginary of community with drones that do not terrorize, surveil, militarize, or exploit humans, but simply help them. They are technological extensions of a future human community in which humans and drones accept each other as equal partners. The movie should not be utilized to support the claim that drones are good. They can do good, but they can also fail. If anything, they have become embodiments of the anthropomorphic machine, which can build a community with humans. In this sense, the movie exhibits a vision of a postanthropocentric community where humans and machines have gained a form of equal agency. This fantasy of drones as companions can open a horizon of reflection to reconsider our relationship with drones, to rethink our interactions, expectations, and connections with them in pandemic times. In doing so, *Rotor DR1* also sets the stage for a further discussion of the pandemic drone sensorium. My next example of pandemic drones is also taken from the realm of amateur film: in this case, drone films of cities under lockdown. In these films, the drone is styled not as a companion but as an aesthetic device that can articulate an empathic perspective.

Pandemic Ghost Town Films and the Drone as Empathy Machine

During lockdowns, domestic drones have been used for investigative reporting, bringing to light heart-wrenching stories—such as the digging of mass graves for unclaimed COVID-19 victims. In March 2020, the photographer George Steinmetz took drone footage of a mass burial site on Hart Island in the Bronx, New York City.[31] Hart Island has been a burial ground for the anonymous dead for centuries: it has been a graveyard for people killed by the Spanish flu epidemic of 1918, by HIV/AIDS, and now by COVID-19. Steinmetz's drone shows a long rectangular pit in the ground, into which the wooden coffins of COVID-19 victims are being placed by a small team of workers. The workers, who are all inmates of the prison on Rikers Island,[32] unload the trucks that bring the dead, and then bury the bodies. Steinmetz's drone footage was widely shared in the news and on social media, illustrating the pandemic death toll beyond the statistics provided in New York governor Andrew Cuomo's daily briefings. The footage was evidence of the desperate conditions in New York City's overcrowded hospitals and morgues, and provided further proof of the inequality of death in the pandemic, many of whose victims are poor, homeless, or without health insurance.

During the filming, the police confiscated Steinmetz's drone and fined him for photographing the island without permission. Steinmetz acknowledged that flight regulations for drones were important, but he maintained that he had flown the drone from an unpopulated parking lot and had not interfered with air traffic.[33] He intended his drone footage to uncover the scale of the pandemic and inform the public about what was happening on Hart Island: "These are humans, and they're basically being treated like they're toxic waste, like they're radioactive."[34] His whole undertaking with this drone film was to call for more empathy. This places the drone in an interesting role precisely because it is so often seen as a technology that evades empathy. Thus Steinmetz's film offers an entry point into my argument that the pandemic drone sensorium can perform a machinic sensing that simulates and stimulates affective experiences of empathy, and in doing so can create a community of empathic agencies.

In a neuropsychological context, *empathy* refers to the ability to feel the suffering of another person and the capacity to change one's perspective to understand that suffering. Grant Bollmer points out, "There are at least eight different major variations on neuropsychological theories of empathy that are not equivalent . . . , though they all tend to position empathy as either the literal experience of another's internal, affective states, or an innate, nonconscious understanding of another at a neurocognitive level, hence differentiating empathy from sympathy or compassion."[35] I am most interested in aesthetic theories of empathy, as I engage with the affective power of artworks where the drone's digital aesthetic sensorium can be seen as an *empathy machine.*

The philosopher Robert Vischer emphasized intuition (*Einfühlung*) as an aesthetic concept in his 1873 work *On the Optical Sense of Form*. Intuition in this sense means that we project ourselves into works of art: we "feel ourselves into" the object we experience.[36] Vischer developed the idea that the soul (*Seele*) is a projection from a viewer onto an object, thus connecting the object to one's own feelings. The art historian Heinrich Wölfflin, who took great inspiration from Vischer, also worked with the notion of *Einfüh-lung* as aesthetic experience, applying it to the reception and interpretation of architecture.[37] His colleague August Schmarsow likewise emphasized *Ein-fühlung*'s immersive, bodily, and three-dimensional aspects in experiences of architecture: "Psychologically the intuited form of three-dimensional space arises through the experiences of our sense of sight, whether or not assisted by other physiological factors. . . . [It] consists of the residues of sensory expe-rience to which the muscular sensations of our body, the sensitivity of our skin, and the structure of our body all contribute."[38] For Vischer, Wölfflin, and Schmarsow, *Einfühlung* was an aesthetic experience based on affective intensity, in which bodily sensation was more important than dispassion-ate, discursive, aesthetic judgment.[39] "Feeling-into" was far more important for the understanding of art than conscious contemplation.[40]

This focus on aesthetic empathic experience also resonates in the work of the art historical thinker and theoretician Aby Warburg. A student of Wölf-flin, Warburg was well versed in nineteenth-century discussions of aesthetic empathy. He developed a theory of iconic images, exploring the formative power of icons to shape and carry collective memory. Warburg's picture atlas *Mnemosyne* (1927) consists of forty wooden panels covered with black cloth, on which are pinned a montage of nearly a thousand (mostly) pho-tographic images of high art (Renaissance, antiquity, biblical art), as well as images of stamps, coins, newspaper clippings, and advertisements.[41] The goal is to demonstrate the continuation of ancient art in European culture and aesthetics. Warburg claims that some of these ancient artworks gain a timeless continuity through their display of human emotions. He is specifi-cally interested in what he calls *pathos formulas*, emotive expressions and ges-tures that resonate in our cultural collective memory and are still circulated today.[42] Pathos formulas are archetypical visual gestures that stem from cultic images of human suffering (for example, Jesus on the cross, the Pietà, or the Virgin Mary as a grieving mother). According to Warburg (and his contempo-rary Erwin Panofsky), these gestures are metahistorical catalysts of a dynamic energy through which the image gains affective intensity and power. As the word *pathos* hints, this universal iconography of suffering can evoke affect and empathy in the viewer. For Warburg, the idea of pathos is tied to the gestures, facial expressions, and postures of the human body.

How does the drone fit into this idea of the pathos formula? Can the drone as an aesthetic medium of disembodiment suggest an embodied view after all? And further, can a drone even shape a community that collectively expresses empathy? At first glance, the drone does not appear to be an aesthetic medium that would evoke *Einfühlung*. Rather, the drone and its sensorium stand for distancing, coldness, and emotional and affective numbness. The machinic vision of the drone seems to be incapable of evoking empathy, particularly if we think of its scopic regime of verticality and militarism. However, I am going to argue that a specific form of drone filming can evoke empathy and shape imaginaries of communities where machines and humans are entangled in an empathic relationship.

Disembodied machine vision has not always been seen as an alienating, inhumane, and oppressive form of perception. In the aesthetic theories of the avant-garde, the machinic vision of a disembodied film camera could act as a catalyst of liberation from the humancentric gaze. According to Antonio Somaini, one finds such moments of liberation in the films and theories created by Dziga Vertov in the 1920s and 1930s: "Traces of the idea of a 'machine vision' can be found in the 'kino-eye' [*kino-glaz*] that captures and reorganizes the visible world through the two operations of optical recording and montage (Vertov 1923), in the 'metal brain' [*cerveau metallique*] of a camera that is 'a nonhuman eye, without memory, without thought' capable of 'escaping the egocentrism of our personal viewpoint.'"[43]

The key phrase here is "escaping the egocentrism of our personal viewpoint." Might there be something in drone filming that can also decenter the human point of view, enabling new forms of empathy and attachment? Harun Farocki's influential theory of the operative image (discussed in chapter 1) suggests not.[44] For Farocki, machine-generated images can execute violence without human agency: disembodied vision can be equivalent to inhumanity, automated warfare, and violence. Nevertheless, as I will show, the COVID-19 pandemic gave rise to a genre of drone films where we can find a nonegocentric viewpoint and the simulation of an empathic observer: flyover drone movies of cities under lockdown, ghost town movies that show the deserted cityscapes of Rome, Boston, New York City, Los Angeles, and others. Often shot by amateurs and posted on YouTube, these films may not be as avant-garde or aesthetically complex as Vertov's work, but they show parallels insofar as the disembodied machinic eye constructs new forms of embodied attachment.[45]

Let us take a look at an example of a COVID-19 ghost town drone movie, *Il Silenzio di Roma* (The Silence of Rome), filmed in April 2020.[46] As is typical of

Figure 7.2
Still from *Il Silenzio di Roma* (2020). © Luigi di Palumbo.

these films, *Il Silenzio di Roma* shows empty streets, squares, parks, buildings, and roads. As the drone flies past and above Rome's most famous sights, the emptiness becomes even more pronounced as one recalls them crowded with people. There are plenty of other films that show such cityscapes, but I have chosen this particular film because here the drone unfolds all its sensorial capacities in a visual, acoustic, rhythmic, and even kinetic-tactile fashion. In *Il Silenzio di Roma*, the drone never wholly takes on the bird's-eye view that would show the city in a distant way, as abstract patterns or a flattened maplike picture. Instead, the drone view is always slightly tilted, and we have an impression of moving with the drone through Rome. The drone becomes our avatar. Indeed, the disembodied drone develops a simulation of the body schema—a prediscursive sensation of our body position in space and time, a nonconscious somatic process of orientation.[47] The film activates this sensor-motoric form of perception via the technology of the drone.

Similarly to the drone races discussed in chapter 2, the viewer is engulfed in a first-person view, gliding with the drone above Rome. As it wanders through the city, the drone does not embody the aggressive all-seeing eye of the vertical, but immerses the viewer's body in the environment. The drone never flies very high; it glides through streets, showing locked-up cafés and empty roads. Nor does it make its way through the city according to a straight route; rather, it loses itself. This is particularly evident when the drone flies around a water fountain several times: it circles it, goes backward, stops, and glides on. The drone acts like an agent of deterritorialization, which for John Johnston is a constructive and liberating aspect of the machine eye that he also sees realized in Vertov's films.[48]

The drone thus immerses the viewer in the city, but the viewer as the embodied drone avatar is not in control of the flight path. It is as if the drone becomes a machinic flaneur. For Walter Benjamin, the flaneur is an important figure of modernity. There is an art to getting lost in a city. As he writes in his *Berlin Childhood around 1900* (*Berliner Kindheit um Neunzehnhundert*): "Not to find one's way around a city does not mean much. But to lose one's way in a city, as one loses one's way in a forest, requires some schooling. Street names must speak to the urban wanderer like the snapping of dry twigs, and little streets in the heart of the city must reflect the times of day, for him, as clearly as a mountain valley."[49] The drone enacts this getting-lost-in-the-city, performing a mimetic, bodily, and sensual exploration of space. In the film, the city is penetrated by the sounds of nature; there is no human-created noise. As its title promises, the film portrays the silence of Rome. With the exception of the ambulance sirens, the only sounds one hears are birdsong and the dripping water of the fountains, a natural musique concrète that has become audible thanks to the lockdown.

This renaturalized cityscape, into which the drone avatar brings us, heightens an affective energy that makes us feel attached and close. In one scene, the drone glides up the Spanish Steps, a site that is usually packed with people but is now completely empty. The drone as telepresence device brings us here: we levitate over this famous space, we are right there. Our telepresence at the Spanish Steps evokes powerful affective energy; the drone film of the site emanates nostalgia, mourning, and a longing for a lifestyle we have lost.

Ada Ackerman describes the aesthetics of *Il Silenzio di Roma* as follows: "Benefitting from exceptional shooting conditions, impossible in normal times, drones recorded the unprecedented situation of stopped cities, stirring feelings of wonder, of melancholia as well as of uncanniness—features shared by 'ruin porn.'"[50] It is true that there is a spectacularization of the sites that COVID-19 created, and that these sentimental, nostalgic views aestheticize a state of political crisis and emergency. But I read these COVID-19 drone cityscapes differently. I see the drone in the film as a sensorial interface that makes it possible to immerse oneself in a space in a prediscursive, visceral fashion, which in turn creates aesthetic conditions for the possibility of empathy. In the film, the drone becomes an aesthetic empathy machine. The somatic and bodily immersion enables the viewer to feel the loss of what we once considered normal: the loss of everyday life, the loss of family members and friends. In this experience of bereavement, the viewer can develop empathy for the humans that are not visible, for the people that have to stay inside their houses, locked away. This moment of empathy, which emanates from the drone, also serves as a trajectory to envision a community: a community of those who have been forced into lockdown or are in mourning for their COVID dead. The drone facilities this type of community; it makes people bond and feel connected.

Immersive technologies such as VGA goggles and other devices have been used with trauma patients, who learn how to manage their trauma through exposure to virtual and augmented realities.[51] Farocki's experimental film *Serious Games* negotiates this technology, which simultaneously inflicts and seeks to heal the wounds of war. Farocki takes a critical view of the military's use of these immersive simulated war scenarios, focusing on the numbing of sensitivity and the healing of trauma. By similarly infusing its viewers with a sense of loss, the drone in *Il Silenzio di Roma* also trains their empathy. The film portrays people only as shadows, vanishing, somewhat lost figures, inhabitants of a space in which they can no longer move around. As Teresa Castro points out, this aesthetics might speak to ecofascist thought, whereby right-wing radicals propagate the idea that humans are a virus of which the planet must be cleansed and purified.[52] But these COVID-19 ghost town films are no invitation to celebrate the extinction of the human race. Rather, I consider the drone and its immersive sensorium to be a training platform on which to rediscover space, rethink the human's position in the world, and reflect on the world we took for granted. The drone camera and its auditory sensorium suggests an empathic and affective entanglement with the space of the city, one that is not marked by instrumental thinking but offers a tender, mimetic exploration of the environment. Thus the drone makes us take on another perspective. No longer is the ego in the center; the camera view absorbs the space without owning it. This is not an idea of purity in the sense that nature is rid of us at last. Rather, the drone reveals the vulnerability of the human. Instead of endorsing extinction, the drone gives us an affective experience of our environment and an opportunity to explore it anew. In this way, this community vision also connects to those discussed in chapter 2, where a form of embodied sensing (such as in the art installation *Cloud Face*) highlighted a noninstrumental relationship between machines, humans, and nature.

Ackerman interprets the moments of beauty in *Il Silenzio di Roma* as manifestations of the grandiose aesthetic sublime.[53] True, the film conveys moments of aesthetic sublimity, but it is not the sublimity of monumentality. Rather, the drone emanates an aesthetic experience that makes us care for places. This is the sublime in Jean-François Lyotard's postmodern sense: it points to the multiplicity and instability of the world.[54]

The drone in *Il Silenzio di Roma* can thus be seen as the catalyst of a pathos formula because it creates aesthetic experiences of heightened affective energy and a trajectory toward feeling the suffering of the other. But this pathos formula no longer works according to Warburg's affective gestures and iconic images. Instead, the drone captures the vanishing shadows of humans; it seizes the human as a vanishing point. Perhaps the disappearing human is the new pathos formula of 2020.

In conclusion, we can state that these COVID-19 drone cityscapes are in some ways reminiscent of the 1920s and 1930s modernist films that portrayed the metropolis in a semidocumentary style. Think of Walter Ruttmann's *Berlin: Symphony of a Great City* (1927), which shows crowds of workers, traffic, railroads, and steaming factories. The machinic eye (camera) and the aerial view immerse the viewer in the feel of modern industry, technological progress, and the power of capital and humanity. But as Caren Kaplan notes, the drones in pandemic cityscapes turn this perspective around: "The coronavirus drone pieces invert this tradition: now the drone is the machine that moves through space in a built environment devoid of crowds, a pattern of buildings (with people presumably inside), and an exterior of roads (with delivery vehicles and a lone bicyclist). They perform an elegy for the stoppage of global capital. . . . The aerial flyover in coronavirus drone imagery now conveys the message that modernity is in peril or decline."[55] The drone as empathy machine makes us reflect on how we interact in and with our worlds, where we position ourselves, and where modernization is going. The drone and its sensorial capacities make us think about our loss. I further explore these issues in the next part of this chapter, where I return to the aesthetics of modernism—this time, the mass ornament—to understand drones as sensing media during the pandemic.

Drone Shows as Mass Ornaments

In drone shows, hundreds of drones illuminate the night sky, forming figures and patterns in the air with various light effects. These digital fireworks serve as public spectacles, for example, at New Year's celebrations or the opening of sports events. At the 2017 Super Bowl, over three hundred drones painted the US flag above the stadium; the giant cover of *Time* magazine's themed issue "The Drone Age" showed over eight hundred drones lighting up the sky. With the advent of the COVID-19 pandemic in 2020, such drone displays increased in number. As social distancing rules were enforced, drone light shows became a riskless form of artistic display during holiday seasons or at sporting events without physical audiences. Drone light shows that explicitly addressed the crisis became especially popular: drone displays would form a giant mask in the sky, or an image of the spiky red COVID-19 virus, or hands ready to receive sanitizer. There were also drone displays that wrote messages, using the sky as a screen for warnings (perhaps reminiscent of the biblical scripture on the wall, *mene, mene, tekel, upharsin,* ["numbered, numbered, weighed, divided"], referring to God's dooming of the kingdom Belshazaar as well as messages of appreciation). At a display in South Korea, hundreds of drones drew hearts and wrote thank you notes in the sky, expressing gratitude to frontline medical staff for their work.

How can we interpret these drone spectacles? Are they simple manifestations of tech companies' profit interests? Are they yet another allegory for the militarization of our pandemic everyday? Similarly to my approach to *Rotor DR1* and COVID-19 ghost town drone movies, I want to find out whether these pandemic aesthetic drone sensoria can provide us with new knowledge about how we see ourselves and our communities. To do so, I will look back through history and delve into the mass culture of the Weimar Republic. This historization of the drone will help me to understand the power of its sensorium and dismantle the image of the drone as a super-new, state-of-the-art technology. It will also open up an experimental space to envision machinocentric empathic communities.

In many ways, today's drone shows resemble the "mass ornament" analyzed by Kracauer almost a hundred years ago.[56] Obviously, Kracauer's famous essay "The Mass Ornament" does not mention drones; it speaks instead of the 1920s phenomenon of mass dance formations, exemplified by the Tiller Girls. This dance troupe staged mass performances in dance halls and sports arenas, and one of its aesthetic goals was the perfect synchronization of the dancers' body movements. Kracauer notes, "One need only glance at the screen to learn that the ornaments are composed of thousands of bodies, sexless bodies in bathing suits. The regularity of their patterns is cheered by the masses, themselves arranged by the stands in tier upon ordered tier."[57]

The dancers form anonymous and abstract clusters. Their disassembled bodies forge "lines and circles like those found in Euclidean geometry,"[58] producing a collective formation that is emptied of individualism. It is notable that Kracauer connects the visual aesthetics of this mass ornament with the aerial view from above: "The ornament resembles aerial photographs of landscapes and cities in that it does not emerge out of the interior of the given conditions but rather appears above them."[59] This aerial photographic quality links the mass ornament to the drone, which similarly provides an aerial perspective defined by abstraction and flattening (see chapter 4). Seen in this light, drone images too can be characterized as ornamental, reiterating Kracauer's modernist aesthetics of the mass ornament, and recalling nonfigurative art.[60] Like the Tiller Girls, the drone show shapes formations with the utmost precision, lines and patterns calculated and defined by linear and algorithmic systems.

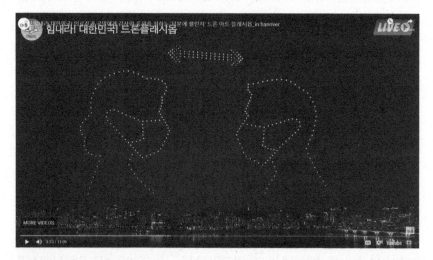

Figure 7.3
South Korea's Ministry of Land Infrastructure and Transport uses hundreds of drones for a spectacular display in support of mask-wearing. © AFP Photo: Ministry of Land, Infrastructure, and Transport.

The drone show displays two human figures with a social distancing arrow above them. The drones display a gigantic collective aesthetic sensorium: the key thing here is not what the drones see, but how they are seen in the sky. The drones use the sky as an artistic screen, illuminating it alongside the stars. It is as if God himself has sent us a text message: remember social distancing! The drones cooperate with each other to form an ornament: human figures composed of a multiplicity of component drones. The drones are singular units that piece together the human figures as giant mobile networks of data processors. As in Kracauer's mass ornament, the human figure here is not an individual; it is instead constituted from what Kracauer calls "building blocks,"[61] fragments and rationalized patterns. Svea Braeunert has interpreted this version of the human figure as a "dividual."[62] The notion of the dividual was introduced by Gilles Deleuze to explain the idea that in modern societies of control, the human individual is endlessly divisible and reducible to data representations via technology. The individual becomes a "dividualized" body of data in the form of "samples, markets, networks or banks."[63] Braeunert suggests that by virtue of this dividualization, drone displays epitomize the imbrication of the military and entertainment. In the words of philosopher Gerald Raunig, "The drone brings death or it brings mail from Amazon, based on algorithmically produced risk or potentiality profiles."[64] The drones in these big displays—like mass ornaments—are connected to militarization because they always follow the logic of humans' dividualization and dehumanization. These processes of dividualization also suggest a community model that is not based on the concept of the individuum, but rather on singularities that can be entwined in nonhuman and human networks. Thus, similarly to Jean-Luc Nancy's theory, it is a model that attempts to think community on the basis of the loss of a common totalizing concept of identity. Community in this case is a configuration of networks, data technology, and entangled subjectivities.

While I am highly aware of the entertainment drone's embeddedness in militarism and capitalism, my aim is to think the mass ornament and the drone show dialectically, using its capitalist–militaristic entanglements as a catalyst of critique. It is true that the dividualization of the human subject, which is apparent in the mass formation of the Tiller Girls, points to the rationalization of capitalism, and the mass ornament is "the aesthetic reflex of the rationality to which the prevailing economic system aspires."[65] According to Kracauer, the legs of the Tiller Girls are the hands of the factory in the Taylorist logic of the assembly line.[66] Moreover, behind every drone light show there is a for-profit tech company, and these companies often also produce military technology. One might think too of the drone fireworks at Joe Biden's presidential inauguration, which was certainly a political statement about the continuation of the drone program.

However, Kracauer did not only debunk the mass ornament as capitalist propaganda, he also used it as an epistemological prism. What intrigues me in Kracauer's thinking is that he sees mass ornaments as having constructive potential to provide knowledge about one's current political situation. He even detects a utopian glimmer in the phenomenon of the mass ornament, which might lead to a historical state of reason: "The process leads directly through the center of the mass ornament, not away from it. It can move forward only when thinking circumscribes nature and produces man as he is constituted by reason. Then society will change."[67] As Kracauer mentions, the ornament embodies a "rational empty form,"[68] which, nevertheless—or for precisely that reason—gives rise to a "thinking that fosters ever greater independence from the natural conditions and thereby creates an intervention of reason."[69] This quote reveals Kracauer's ties to the negative dialectics of the Frankfurt School and its critique of the Enlightenment discourse of reason. He hopes that the mass ornament's disenchanting powers will create room for a new form of reason and rationality, one freed from any naturalized interrelations with capitalism and militarism.

I consider Kracauer's ideas about the mass ornament to be an invitation to rethink the drone display beyond its immediate bonds to capitalism and militarism. Instead, we can consider the drone display as an aesthetic experiment in conceptualizing the self and our communities beyond human–machine dichotomies in pandemic times. No less than two decades ago, Jay David Bolter and Richard Grusin showed how the digitalization of society shapes a conception of the self that is based on multiple layers and hypermediality: "Even in the case of virtual reality, which tries to construct a purely transparent version of the digital self, there is a constant interplay between transparency and hypermediacy. . . . In such cases, the self is expressed in the very multiplicity and fragmentation of the windowed style."[70] The self is fragmented due to its constant digital remediation, a "networked self" that consists of many digital reflections, forms, and layers. The pandemic has in many ways intensified this networked form of the self, and its digital multiplicities have expanded globally as many of us interact with the world more and more digitally. The sheer quantity of connections via videoconferencing platforms, such as Zoom, Microsoft Teams, and Skype, mean there are large data quantities of ourselves in platforms, archives, and data storage facilities. Yvonne Zimmermann has discussed the self-reflexive, multiple mirror process that Zoom conferences evoke: we want to connect to others, but we also constantly monitor and observe our various digital selves.[71] Like the calculated elements of the drone show mass ornament, our sense of self amid all this pandemic videoconferencing also seems to be broken into multiple layers, manifested in rectangles, lines, split screens, waiting rooms, and breakout rooms.

The pandemic drone ornament thus allegorizes a model of subjectivity in which the subject and the self are no longer monolithic or static. Subjectivity now consists of many heterogeneous datafied layers that constantly shift and morph in different places and time zones. Similarly, with AI formations, such as Siri and Alexa, we can interact simultaneously across temporal and spatial borders. I am not suggesting that this is a desirable state, but I read the fragmentation of the individual in the drone ornament beyond the military-dividualization pattern. The mass ornament and the drone display inspire me to think subjectivity in more fluid, relational ways, and this provides me with the opportunity to embrace technology as a nonobject in a nonhuman–human community.

According to Kracauer, the mass ornament and its rationality of lines, forms, and calculus extract the human from its bond to nature. The mass ornament calls a halt to the mythologization of nature and human communities as organic forms. Instead, it disenchants communities: "The bearer of the ornaments are the masses, and not the people [Volk]."[72] For Kracauer, the Tiller Girls are a collective that has freed itself from societal models of essentialist communities that follow organic processes. The Tiller Girls have overcome the power and domination of nature by becoming a technological rationality; thus, in an obscure way, they have made way for a form of reason beyond political instrumentalization. I suggest that in similar ways, drone displays can also be seen as a cipher for a nonessentialist form of community, a form I connect to Nancy's idea of nonidentical communities. This community can also connect to models of the swarm and the multitude (see chapter 6), because it is based not on the individual subject but on a collectivity that embraces human and nonhuman matters. Again, I do not intend to fetishize or idealize the new pandemic virtual communities that the drone ornaments illustrate. But I do read them as prisms that give insights into how the pandemic and its digitalization are changing our configurations of ourselves and our modes of being together.

Pandemic Communities

This chapter shows that the drone plays a central role in the medialization of the pandemic. The drone and its sensoria can intervene in the pandemic by spraying disinfectants, monitoring crowds, and detecting temperatures. Further, artists use drones as aesthetic media and provide aesthetic imaginaries of the pandemic. In all these aesthetic sensoria, the pandemic drone negotiates models of communities. As I have discussed in this book, the drone is a technology that enforces social sorting and surveillance. In this way, the drone can act as a biopolitical device of social segregation, sorting the infected from the noninfected, separating the healthy from the unhealthy. Many researchers have pointed to infringements of civil liberties and privacy in this respect, and they see the drone as a powerful technology that enforces police and state surveillance. The drone in this sense is a technology of control embedded in the power discourses of militarism.

But besides these violent, inhumane, and military forms of community shaping, I have shown that the pandemic drone also has the power to imagine communities that are not defined by those criteria. In artworks in particular, the drone can suggest visions of communities where remote sensing technology does not suppress but empowers the human agent. The pandemic drone as an aesthetic medium can experiment with new forms of social bonding, creating alliances between the human and the machine. The speculative imaginary of *Rotor DR1* suggests the drone as a companion to cope with the devastation of a pandemic. The film portrays a human community that has lost everything but has learned to relate to machines. These drones are not glorified as saviors; they simply coexist with humans in equal ways. This projection of an emotional and affective community between humans and drones is also present in the COVID-19 ghost town movies, in which the drone evolves into a medium that provides an empathic perspective. The drone mass ornament gives insights into how we today might imagine subjectivity and communities where the human subject/individual is no longer the exclusive center. All of the forms of communities discussed in this chapter are based on a different notion of subjectivity, one that can relate to nonhuman objects (places, landscapes, architecture, machines, networks). In some of the artistic imaginaries that the pandemic drone creates, nonhuman-centered visions of communities can flare up. These artworks do not discuss the political implications of drone–human bonding during the pandemic; their visions remain obscure, opaque, even paradoxical. Nevertheless, they can be seen as clues to how the pandemic is changing our sense of community. Perhaps these aesthetic drone imaginaries of pandemic communities articulate how unlimited individualism and self-realization have come under siege. States have become stronger; restrictions and regulations determine our daily lives; we are told to act for others' sake, to protect them. I am far from celebrating any solidarity that the pandemic has supposedly brought, as it has become clear that vaccines are distributed in nonsolidaristic ways, according to the principles of profit and capital. The artworks discussed in this chapter epitomize an aesthetic Richter scale, registering the potential of new forms of communities and bonding between human and machinic worlds.

Conclusion

Mnemodrone is an art installation about a drone that stores personal memories and speaks with people about them. It contains the crux of what I want to say in this book: drones can embody a medium of futurity. Let me explain this. The word *mnemonic* refers to memory and relates to the Greek goddess Mnemosyne. Ancient Greek philosophers cultivated mnemotechnics as a strategy to memorize public speeches. The artwork *Mnemodrone* is also a prop for remembering—not speeches, but personal memories. In 2014, *Mnemodrone* sat (among other places) on the ground in a park in Brooklyn. Visitors gathered around it, as if around a storyteller (recalling the myth of storytelling as discussed with Jean Luc Nancy's concept of community in chapter 1), and shared their personal memories with the drone. They talked about their first loves, their childhoods, and other personal experiences. The drone listened patiently and stored all of these remembrances.[1] *Mnemodrone* embodies an archive, a database, and a storehouse of human experiences: it is a collective *lieu de mémoire*, that is, a subjective and affective form of memory based in a spatial setting.[2]

But *Mnemodrone* is not just an apparatus that stores, conserves, and seals the past in a time capsule. It also opens up a dimension of futurity. As Walter Benjamin suggests, a glance backward into the past can evoke a glimpse into the future.[3] *Mnemodrone*'s stored memories point toward future interactions, communities, and modes of human–machinic attachments. Humans and machines do not work against each other; rather, they embrace each other. *Mnemodrone* engages in dialogue, interacts, retells, and even makes new stories. It can do so because its memory data form the foundation of a database, which is then processed by an artificial intelligence (AI) machine learning program. The drone has learned to talk back. Through this AI

Figure C.1
Image from Emilio Vavarella and Daniel Belquer's postanthropocentric art project
Mnemodrone (2014–). Mixed media, multiple locations. Photo courtesy the artists.
© Emilio Vavarella and Daniel Belquer.

program, the drone can converse about personal memory, beauty, and existential questions of life and death. Such dialogues with the drone suggest meaningful interaction, since *Mnemodrone* produces the answers logically and (it seems) intuitively. It even seems to have an emotional and philosophical grasp of the world.

This ability of *Mnemodrone* to communicate, respond, and interact suggests the vision of community between machines and (drone) technology that this book has attempted to trace: a community where machines and humans form a relational and mutual assemblage that is not based on domination, instrumentalization, or hierarchies. This vision of community in *Mnemodrone* has a dimension of futurity because it suggests a model of social interaction and community-building. The collecting of past memories raises questions about how humans interact with a drone, what their concerns and ambiguities are, how they trust the drone, and how humans tend to anthropomorphize nonhuman agents. *Mnemodrone* gains this community-building power by performing as if it had human consciousness. It reflects the current possibilities of AI as an approximation of human consciousness, and it foreshadows what might be possible in the near future. Emilio Vavarella and Daniel Belquer, the creators of *Mnemodrone*, describe the project as postanthropocentric because it engages with "machinocentric creativity" and AI, decentering the human in communicative, social, and aesthetic processes.[4]

This power of drone futuring is what this book has sought to unearth. Aesthetic drone sensoria and their respective imaginaries of communities can construct visions of the future. They can provide us with speculative scenarios regarding how humans might live with machines, and how we might relate to remote sensing technology beyond its rationalized and militaristic instrumentalization. This type of futuring has nothing to do with preemption. It is a modern type of futuring based on dynamism, change, and open-endedness. But I am not suggesting that drones follow a higher telos of history, striving for the perfectibility of technology. I simply dare to infuse the drone with a small dose of philosophical playfulness. Specifically in the realm of the arts, drones can make us reflect on our relationship with intelligent machines in new ways and can potentially shape our future in ways that will be beneficial for the planet.

This view of drones is unusual in the humanities. Typically, drones are regarded as technological devices that nullify the future by determining it before it even begins. They are rather premodern that way.[5] The drone and its capacity for constant surveillance can predict future crimes. In combination with AI and predictive algorithms, the military drone can foresee terror operations and insurgencies.[6] In these preemptive operations, the drone annihilates the future, because it seems to know what will happen in advance: technological determinism and essentialism. Intervening in the contingencies of life, predicting fates before they are enacted, is a dangerous business. In the film *Minority Report* (2002), cogs, machines, and humans detect premeditations of crimes, and this leads to the arrest of innocent people. These "offenders" have not (yet) committed crimes, and they might have been about to change their minds. The very essence of the future, namely its unpredictability and contingency, is calculated away. We all know that this type of preemption is no longer fiction: the drone may be *the* technology to execute preemptive war strikes or precrime actions. As a military drone pilot once admitted, "Sometimes I felt like a God hurling thunderbolts from afar."[7] Often, discourses that nestle the drone in a scopic regime of omnivoyance accentuate its Olympian powers.[8] As Brian Massumi notes, impersonating the view of the drone, "We, the preemptors, are the producers of your world. Get used to it."[9] A godlike creature, the drone can effortlessly cross the boundaries of space: it can be everywhere. The drone's spatial mastery is deeply ingrained in the cultural and political narrative of the aerial view. From God's eye to the panopticon to the drone, the aerial embodies the controlling eye.

It is true that drones can rule space, like wrathful gods. But they often also fail, and they can do many other things. My book has shown that drones do not always take an overview; their vision can be clouded, blurry, and uncertain. They are not always about distance but can create intimate and personal relational spaces. They can sense space volumetrically and atmospherically, which in turn proliferates new relations and connections to our sensorial earth environments. They also have a power to open up the future, making it permeable to visionary imaginaries of communities.

The chapters of this book have illustrated how the synesthetic aesthetic sensorium of the drone embodies an affective matrix that experiments with new visions of collectives and alternative forms of bonding between humans and machines. Part II, on the drone and the body, demonstrated how the drone sensorium, with its embodied and telepresent modes of sensing, creates entangled relations between the body and the drone, and how these drone technobodies suggest cyborg assemblages that overcome human-centered forms of communities. The drone no longer represents a detached surgical apparatus, but rather brings people together, creating intimacy and closeness.

But the drone sensorium does not only unfold its futuring powers in relation to the body. Part III, on drones and the earth, presented us with new ways to connect to the earth via the drone. The drone as eco-medium attaches us more closely to the earth by facilitating an attitude of care, relating us to the earth through the lens of the planetary and the Anthropocene. Thus the drone sensorium creates trajectories for postcarbon futures, making us reflect on the current state of our planet, its ecological condition, and the changes to its climate. Paul Cureton sees drone futuring as vital to save the planet: "UAS' [unmanned aircraft systems'] public perception and the highlighting of drones as a social good, which can assist many areas of everyday life in the natural and built environment, will develop acceptance. . . . UAS will then also provide a powerful medium for monitoring climate, place, space and our lives."[10] I am not interested in deciding whether such drones are good or bad per se, but in what we can learn from their aesthetic imaginaries.

These earth drones do not separate us from the earth but enmesh us with its materiality.

In bringing us closer to the earth, the drone also engages us with what we see as nonhuman. Part IV, on the drone and the nonhuman, demonstrated that the aesthetic drone sensorium can imagine communities that integrate a zoological, nonhuman, swarmlike dimension. Here, the swarm is not seen a priori as the bestial dehumanized "other," but offers a trajectory for thinking communities as multitudes.

Many of the diverse aesthetic imaginaries of drone communities that I have discussed in this book have utopian glimmers at their core. Some of them show communities where hierarchies between machines and humans are flattened, and where the drone and the human are thought together in a creative embracing assemblage. Some suggest planetary communities that seek to overcome the territorialization of the earth by exclusion, colonialism, and imperialism. I have traced community visions that connect to Jean-Luc Nancy's and Rosi Braidotti's ideas about communities, in which notions of identity, origin, and authenticity no longer dominate. I wanted to show that the multisensoriality of the drone can shape a template for imagining communities that are defined neither by the "collective singular"[11] nor by the violent exclusionary networks of predictive algorithms. The aesthetic drone imaginaries of communities that I have harvested from drone art reveal the life-affirming and vital potentials of the remote and machinic drone sensorium, and experiment with community visions that arise from thought surrounding difference, otherness, and posthumanism.

This futuring with the drone makes us rethink our relation to the drone's remote sensing technology. The drones in this book are much more than flying objects with sensors and computers inside. The artworks I have discussed show that drones are epistemological engines: they shape the ways we think, interact with each other, and create worlds with technology. We can be closely entangled with drone technology; it can become like us. Following Benjamin once more, one might say that my book compares the drone to a natural "organ." Benjamin noted that in the face of war, "society was not mature enough to make technology its organ."[12] Wars are still being fought today, but, perhaps in light of our planetary vulnerability, we can begin to see technology as *one of us.* Benjamin connects his ideas about technology to the notion of "organ projection" developed in Ernst Kapp's 1877 text *Elements of a Philosophy of Technology (Grundlinien einer Philosophie der*

Technik).[13] Kapp claims that all technical artifacts are extensions of human organs (hammer as fist, binocular as eye). This relationship between a technical object and an organ speaks to an anthropological conception of technology: technology develops in sync with the human and its anthropomorphic body. But instead of reading this organ theory as a narrative of human mastery over technology, I consider Benjamin's and Kapp's organ metaphor as an invitation to understand technology as a type of organism, a vital cybernetic system that regulates itself and is closely entwined with human and nonhuman environments. This perspective on technology also recalls Yuk Hui's ecology of machines, which diminishes the ontological distance between the environment of technological objects and the subject.[14] In this book, I have also considered drone technology as an organ-assemblage, seeing the human and the technological as milieus that are no longer divided by ontological difference but closely intertwined.[15]

But to see the drone as an organ and excavate its potential futuring power is to play with fire, even when one remains within the realms of aesthetics, fiction, the imaginary, and speculative fantasy. Drones come from a violent family, and many of the scenarios one can imagine with them are bleak. That is also why this book has devoted many passages to the drones' dark side. Although the violent (and dystopian) drone imaginaries that I have discussed in relation to facial recognition and embodied sensing do not adhere to the scopic regime, they can still execute violence, precisely because of their decentralized and networked configuration. In these imaginaries, the drone loses its alternative power of futuring and instead shapes dystopic and dehumanized imaginaries of communities.

The realm of aesthetics is a privileged one in which to confront these paradoxicalities of the drone and the duality of its vital/deadly imaginaries of technosensual communities. Art does not care about contradictions. We should bear that in mind when we interpret and discuss drones. As Melvin Kranzberg notes, "Technology is neither good nor bad; nor is it neutral."[16] My readings of drone artworks do not engage in a normative discourse about what we ought to do. Rather, the aesthetic imaginaries of the drone function as prisms through which we can learn about technology and its discursive powers. They should inspire us to rethink drones, to see them critically as well as to find their constructive potentials. We do not always have to resolve the contradictions of technology and categorize it as good or bad. Aesthetics offers a platform on which we can approach the Faustian tensions of the drone. Whether they are from heaven or hell, or from in-between, or from nothing, the point is to extract knowledge from these drone artworks about our relationships with technology. These aesthetic drone imaginaries are seismographs of future applications of technologies. It is time for all of us—not just artists and humanities scholars—to begin to decipher their epistemological palimpsests.

Afterword

My general undertaking to engage with the drone must also be seen within a larger framework: my efforts to show that the humanities are an important player in the field of technology studies. The humanities have a long tradition of researching technology. There is research on aesthetic representations of technology in visual arts and literature, as well as whole fields, such as the history of technology, the philosophy of technology, cultural studies, media studies, design, and the digital humanities. This list is by no means exhaustive, and in this afterword I am not able to map out these different traditions. But there is perhaps one common denominator that connects some of the different humanities approaches to technology: the idea of techné.

Techné contains the Greek linguistic root *tek*, which can mean to build, do carpentry, make something from wood, or construct. In Homer's *Iliad*, techné is connected to the skill of ship construction, to Hector's craft and dexterity with an ax in the process of building a boat. The Greek Sophists also have an interest in the notion of techné, which is understood very much in light of the technification of daily life, and which occupies a place in rhetoric, art, politics, medicine, music, and military strategy. Aristotle defines techné as the craft to create and to generate something that did not exist before. Thus techné is different than nature (*physis*), which according to Aristotle has movement in itself and is self-acting. In contrast, techné is the product of the human, the engineer or the artist; it is the product of the human capacity to bring something forth into the world.

Martin Heidegger takes up the notion of techné in his 1954 essay "The Question Concerning Technology" ("Die Frage nach der Technik"). He states that "the essence of technology is by no means anything technological."[1] Similarly to Aristotle, Heidegger understands techné as a human activity to bring something forth into the world. Indeed, Heidegger closely interrelates the field of art with technology because both entail the human capacity to create something imaginary. Heidegger says, "Techné belongs to bringing-forth, to poesis, it is something poetic."[2] Like poetry, techné creates possible worlds and potential actualities. Techné is like a language with which we can understand, project, and shape the world hermeneutically.

For Heidegger, technology as techné is not about an apparatus with which we can fix problems. Technology is not an instrument. For Heidegger, homo faber—the making human as the progress-oriented, rational, and calculating engineer—is blind. Max Frisch's 1957 novel *Homo Faber*, a work surely influenced by Heidegger's musings on technology, shows the collapse of Walter Faber's world. Not only does Faber's airplane have to make an emergency landing but he also—by chance and unknowingly—falls in love with his own daughter, Sabeth. Faber eventually meets Hannah, his estranged former lover and the mother of his daughter. During all this, he learns about his own blindness to the world, which is rooted in his rational and instrumental conception of technology.

Heidegger likewise does not see technology as a tool with which to master the world. Rather, his idea of techné makes technology into a philosophical-poetic activity that gives us access to the truth of being. Via techné, Heidegger aims to uncover a premodern understanding of technology, which for him should open a way to engage with his ontology of being and his critique of metaphysics. This is very much the point at which I separate from Heidegger, who essentializes techné as a vehicle for his conception of truth. In doing so, he problematically romanticizes preindustrial technology, fostering a nostalgic critique of civilization.

Nevertheless, I believe that Heidegger's ideas about techné can be productive for us to think about why technology is a humanities subject. Heidegger's philosophy demonstrates that technology, like art, has the capacity to project fictional worlds. The computer program of an engineer, like the pen or brush of an artist, can project hypothetical "as-if-like" worlds. The humanities, with their analytical skills, can observe these technological imaginaries, interpret them, frame them, and translate them. We as humanists have the power of the imagination to discuss technology under the auspices of counterfactuality: What could and should technology be like? And we have the ability to detect narratives about technology: Are we engaged in a technological love story, or a tragedy? Take the Greek god Prometheus, for example, whose narrative seems to be a technological romance. He gave fire to humans; he embodies innovation, optimism, emancipation, and progress. But the romance has a flip side: with the gruesome punishments he had to endure, Prometheus certainly paid a price for his technological optimism. Prometheus as the avatar of the self-made man is also a tragic figure, symbolizing alienation, the rift between nature and humanity, and the incessant, self-destructive Faustian drive to compete. I believe a key competence of the humanities is to analyze such narratives about technology, to think about alternate plots (comedy, farce, or melodrama, for example), and to reflect on them critically in light of our current lifeworlds. Indeed, I believe this competence is truly needed in a world that often deploys for-profit technofix solutions to societal problems.

There are a couple aspects of techné that I consider important to the field of humanistic technology studies: culture techniques (*Kulturtechniken*), aesthetics, and critique. Techné as poesis implies a close entwinement between culture and technology. This relationship has often been characterized as antagonistic and alienating: culture is threatened by technology, and technology is not part of culture because it is simply not beautiful enough. However, in recent decades the humanities have discussed the interdependence of culture and technology by pointing to the notion of culture techniques. Inspired by the semantics of *agriculture*—*cultura* as the tilling of a field—culture techniques are concepts for handling life situations (making fire, language, maps, images).[3]

Technology as techné shares family resemblances with the arts, poetry, and fiction. Technology, however, also plays an important role in a neighboring discipline, namely aesthetics. This is the branch of philosophy that deals with questions of beauty and taste, as well as (and most importantly here) with the nature of sensual knowledge. The concept of aesthetics as sensual knowledge can be traced from Aristotle down to Alexander Gottlieb von Baumgarten, Christian Wolff, and Immanuel Kant. The school around the German philosopher Gernot Böhme has revitalized this idea of aesthetics as sensual knowledge, specifically in light of technology.[4] Technological inventions demand their own technosensualities and experiences, which is the key topic of this book on drones.

One must be aware, however, that the idea of aesthetics as an epistemic discipline of sensual knowledge also limits the autonomy of art. Theories about sensing and aesthetic judgment are not art. Art does not need a discipline; it does not have to adhere to sense-making rules. It is precisely this freedom that makes art so special. Art offers the conditions in which to observe the world differently, that is, noninstrumentally, affectively, and nondiscursively. This power of art is absolutely vital when we engage with technology in the humanities. Artistic interpretations of technology (whether in sci-fi novels, experimental film, or visual art) can reflect technology's vulnerability, uncertainty, and fallibility. In artists' work, representations of technology evoke dissensus: artworks can negotiate, criticize, ironize, queer, celebrate, and make technology strange.

This view of critique requires me to engage with technology by moving beyond established thought traditions and exploring new ways of thinking about technology and humans. Is techné something of which only humans are capable? Is there a way of reading techné in a posthuman context? Artificial intelligence, autonomous machines, smart technologies, and algorithms are capable of poesis: they are able to make the world, and to create meaning. I consider it of the utmost importance that humanists—as experts in deciphering and analyzing the production of meaning—should engage with these new abilities of machines, which produce new forms of intelligence. We need not only to reevaluate homo faber but also to embrace thinking about technology in new materialist constellations. We need to evaluate our human control seat amid technology in light of cyborgs, machinic sensing, smart algorithms, humanoid robots—and of course, drones.

Notes

Introduction

1. I owe this reference to Agi Haines to the work of Claudette Lauzon, who discusses the drone as a blob, thereby highlighting its violent and military implications. Connecting the blob to horror movies, she calls drone war "blobular" war to refer to both its imprecision and its dehumanizing optics. Claudette Lauzon, "Stranger Things: A Techno-Bestiary of Drones in Art and War," in *Drone Imaginaries: The Remote Power of Vision*, ed. Andreas Immanuel Graae and Kathrin Maurer (Manchester: Manchester University Press, 2021), 180–202.

2. Lauzon, "Stranger Things," 180.

3. For further reading on the sensorial approach to the drone, see Daniela Agostinho, Kathrin Maurer, and Kristin Veel, eds., "The Sensorial Experience of the Drone," themed issue, *The Senses and Society* 15, no. 3 (2020).

4. Sarah Tuck has critiqued the vertical paradigm of the drone by questioning Hito Steyerl's discussion of the vertical as a "proxy perspective that projects delusions of stability, safety, and extreme mastery," thereby also disrupting Stephen Graham's account of the "insidious militarisation of everyday life." See Sarah Tuck, "Drone Vision and Protest," *Photographies* 11, nos. 2–3 (2018): 172; Hito Steyerl, "In Free Fall: A Thought Experiment on Vertical Perspective," *e-flux journal* 24 (2011): 8; Stephen Graham, *Vertical: The City from Satellites to Bunkers* (London: Verso, 2016). See also Sarah Tuck, "Drone Alliances," in *Fragmentation of the Photographic Image in the Digital Age*, ed. Daniel Rubenstein (New York: Routledge, 2020), 73–79.

5. E.g., Kyle Grayson and Jocelyn Mawdsley, "Scopic Regimes and the Visual Turn in International Relations: Seeing World Politics through the Drone," *European Journal of International Relations* 25, no. 2 (2019): 431–457.

6. There is research that approaches military drones and their political impact on social formations beyond such dualisms, and my own scholarship in this book builds on this work. E.g., Cara Daggett, "Drone Disorientations: How 'Unmanned' Weapons Queer the Experience of Killing in War," *International Feminist Journal*

of Politics 17, no. 3 (2015): 361–379; Katherine Chandler, *Unmanning: How Humans, Machines and Media Perform Drone Warfare* (New Brunswick: Rutgers University Press, 2020); Louise Amoore, "Algorithmic War: Everyday Geographies of the War on Terror," *Antipode* 41, no. 1 (2009): 49–69. In the aesthetic context, I have been guided by Thomas Stubblefield's *Drone Art: The Everywhere War as Medium* (Oakland: University of California Press, 2020) and Ronak K. Kapadia's *Insurgent Aesthetics: Security and the Queer Life of the Forever War* (Durham, NC: Duke University Press, 2019).

7. Here I draw on the notion of "fluid surveillance" articulated by Zygmunt Bauman and David Lyon in *Liquid Surveillance: A Conversation* (Hoboken: John Wiley & Sons, 2013), and on the network theory developed by Alexander R. Galloway and Eugene Thacker in *The Exploit: A Theory of Networks* (Minneapolis: University of Minnesota Press, 2013).

8. Charles Taylor, *Modern Social Imaginaries* (Durham, NC: Duke University Press, 2003).

9. On the role of domestic drones with respect to social movements, I have been guided by Austin Choi-Fitzpatrick, *The Good Drone: How Social Movements Democratize Surveillance* (Cambridge, MA: MIT Press, 2020).

10. For this approach to communities, see Benedict Anderson, *Imagined Communities: Reflections on the Origin and Spread of Nationalism* (London: Verso, 2006).

11. There are scholars in science and technology studies who critically reflect on the humanitarian purposes of civilian drones, discussing privacy issues, ethical questions, regulation, and perception. See Kristin Bergtora Sandvik and Maria Gabrielsen Jumbert, eds., *The Good Drone* (New York: Routledge, 2016).

12. Ole B. Jensen, "New 'Foucauldian Boomerangs': Drones and Urban Surveillance," *Surveillance & Society* 14, no. 1 (2016): 20–33.

13. Friedrich Kittler, *Gramophone, Film, Typewriter* (Stanford: Stanford University Press, 1999); see also Paul Virilio, *War and Cinema: The Logistics of Perception* (London: Verso, 1989). For research that considers the drone (including in its domestic form) as entrenched in militarism, see Michael Richardson, "Drone Cultures: Encounters with Everyday Militarisms," *Continuum* 34, no. 6 (2020): 858–869.

14. To name just a few exemplary scholars, Maximilian Jablonowski, "Dronie Citizenship?" in *Selfie Citizenship*, ed. Adi Kuntsman (London: Palgrave Macmillan, 2017), 97–106; Maximilian Jablonowski, "Der Nomos des Vertikalen: Zur Ortung und Ordnung ziviler Drohnen," in *Ordnungen in Alltag und Gesellschaft: Empirisch-kulturwissenschaftliche Perspektiven*, ed. Stefan Groth and Linda Mülli (Würzburg: Königshausen & Neumann, 2019), 77–92; Anna Jackman, "Sensing," Society for Cultural Anthropology, *Theorizing the Contemporary, Fieldsights* (blog), June 27, 2017, https://culanth.org/fieldsights/sensing; Bradley L. Garret and Anthony McCosker, "Non-human Sensing: New Methodologies for the Drone Assemblage," in *Refiguring Techniques in Digital Visual*

Research, ed. Edgar Gómez Cruz, Shanti Sumartojo, and Sarah Pink (London: Palgrave Macmillan, 2017), 13–23.

15. I owe this formulation and precise insight into my book's aim to one of my reviewers, who made extremely helpful comments about my work.

16. Ole B. Jensen uses this term in conjunction with Don Ihde's phenomenological approach to technology. Ole B. Jensen, "Thinking with the Drone: Visual Lessons in Aerial and Volumetric Thinking," *Visual Studies* 35, no. 5 (2020): 419.

17. Alexander von Gottlieb Baumgarten, *Ästhetik*, ed. Dagmar Mirbach (Hamburg: Felix Meiner, 2007), first published 1750.

18. For a connection between Baumgarten's idea of aesthetic training, warfare, and military forms of gaming, see Anders Engberg-Pedersen, "Technologies of Experience: Harun Farocki's *Serious Games* and Military Aesthetics," *boundary 2* 44, no. 4 (2017): 155–178.

19. Alexander Gottlieb von Baumgarten, "Brief 7," in *Philosophische Briefe des Aletheophilus* (Frankfurt/Leipzig, 1741), quoted in *Historisches Wörterbuch der Philosophie*, vol. 1, ed. Joachim Ritter, Karlfried Gründer, and Gottfried Gabriel (Basel: Schwabe, 2005), 557.

20. Baumgarten, "Brief 7," quoted in Ritter, Gründer, and Gabriel, *Historisches Wörterbuch*, 557.

21. Amid this extensive research field, the work of Beatrice M. Fazi can be seen as an example of the discourse on aesthetics. Beatrice M. Fazi, "Digital Aesthetics: The Discrete and the Continuous," *Theory, Culture & Society* 36, no. 1 (2019): 3–26.

22. Caroline A. Jones, "Introduction," in *Sensorium: Embodied Experience, Technology, and Contemporary Art*, ed. Caroline A. Jones (Cambridge, MA: MIT Press, 2006), 1. See also Caroline A. Jones, "The Mediated Sensorium," in Jones, *Sensorium*, 5–49.

23. Jones, "The Mediated Sensorium."

24. Jones, "The Mediated Sensorium," 2.

25. Jones, "The Mediated Sensorium."

26. Peter Adey, "Making the Drone Strange: The Politics, Aesthetics and Surrealism of Levitation," *Geographica Helvetica* 71, no. 4 (2016): 319–329.

27. For aesthetic approaches to the military drone, see e.g., Kapadia, *Insurgent Aesthetics*; Stubblefield, *Drone Art*.

28. Atef Abu Saif, *The Drone Eats with Me: Diaries from a City under Fire* (Manchester: Comma Press, 2015); George Brant, *Grounded* (London: Oberon Books, 2013).

29. Tom Hillenbrand, *Drohnenland* (Cologne: Kiepenheuer and Witsch, 2015).

30. I owe this idea of historizing the drone to Caren Kaplan, whose research has revealed the connection between hot-air balloons and drones in a military context. Caren Kaplan, "Balloon Geography: The Emotion of Motion in Aerostatic Wartime," in *Aerial Aftermaths: Wartime from Above* (Durham, NC: Duke University Press, 2017), 68–103.

31. Ernst Jünger, *The Glass Bees* (New York: Noonday Press, 1960), first published 1957 in German.

32. For a great discussion and overview of drones and art in a military context, see Svea Braeunert and Meredith Malone, eds., *To See without Being Seen: Contemporary Art and Drone Warfare* (Chicago: University of Chicago Press, 2016).

33. Kapadia's *Insurgent Aesthetics* is groundbreaking here, as he discusses drone art from non-Western backgrounds.

34. Astrid Gynnild, "The Robot Eyewitness: Extending Visual Journalism through Drone Surveillance," *Digital Journalism* 2, no. 3 (2014): 334–343.

35. Ariella Azoulay, *Civil Imagination: A Political Ontology of Photography* (London: Verso, 2012). See also Hito Steyerl, "In Defense of the Poor Image," *e-flux journal* 11 (2009), http://www.e-flux.com/journal/view/94.

36. David E. Nye, *American Technological Sublime* (Cambridge, MA: MIT Press, 1996).

37. Here I have been inspired by Sianne Ngai's work on aesthetic categories that lie outside the traditional concepts of the beautiful and the sublime. Sianne Ngai, *Ugly Feelings* (Cambridge, MA: Harvard University Press, 2009).

38. Walter Benjamin, *The Work of Art in the Age of Its Technological Reproducibility, and Other Writings on Media*, ed. Michael W. Jennings, Brigid Doherty, and Thomas Y. Levin (Cambridge, MA: The Belknap Press of Harvard University Press, 2008), 42.

39. I owe this idea of the drone as a reversible image to Svea Braeunert, "Disappearing, Appearing, and Reappearing: Imaging the Human Body in Drone Warfare," in Graae and Maurer, *Drone Imaginaries*, 91–109.

40. Richardson, "Drone Cultures," 858–869.

41. Donna Haraway, *Manifestly Haraway* (Minneapolis: University of Minnesota Press, 2016), 9.

42. Michel Foucault, *Discipline and Punish: The Birth of the Prison* (New York: Pantheon, 1977), 25.

43. Judith Butler further refines this idea by arguing that bodies are politicized in specific cultural settings. Judith Butler, *Gender Trouble: Feminism and the Subversion of Identity* (London: Routledge, 1990); Judith Butler, *Bodies That Matter: On the Discursive Limits of Sex* (London: Routledge, 1993).

44. Kim Stanley Robinson, *The Ministry for the Future* (New York: Orbit Books, 2020).

45. Yuk Hui, "Machine and Ecology," *Angelaki* 25, no. 4 (2020): 56.

46. Hui, "Machine and Ecology."

47. For a critique of the flattening cartographic paradigm in surveillance studies and drones, see Francisco Klauser, *Surveillance and Space* (London: Sage, 2016); Francisco Klauser, "Looking Upwards: Drones and the Social Appropriation of Airspace," in *Game of Drones: Of Unmanned Aerial Vehicles*, ed. Claudia Emmert et al. (Berlin: Neofelis, 2020), 164–173.

48. I owe this insight to Andreas Immanuel Graae, "Swarm of Steel: Insects, Drones and Swarming in Ernst Jünger's *The Glass Bees*," in Graae and Maurer, *Drone Imaginaries*, 149–166.

49. Stubblefield, *Drone Art*, 79–103.

50. Caren Kaplan, "Drone-o-Rama: Troubling the Temporal and Spatial Logics of Distance Warfare," in *Life in the Age of Drone Warfare*, ed. Caren Kaplan and Lisa Parks (Durham, NC: Duke University Press, 2017), 161–177.

Chapter 1

1. Arthur Michel Holland, *Eyes in the Sky: The Secret Rise of Gorgon Stare and How It Will Watch Us All* (Boston, MA: Houghton Mifflin Harcourt, 2019).

2. This attribution of hypervisual powers to the military drone is exemplified in the following: Grégoire Chamayou, *Drone Theory* (London: Penguin, 2015); Peter Adey, Mark Whitehead, and Alison J. Williams, "Introduction: Air-Target Distance, Reach and the Politics of Verticality," *Theory, Culture & Society* 27, no. 6 (2012): 173–187; Hugh Gusterson, *Drone: Remote Control Warfare* (Cambridge, MA.: MIT Press, 2015); Ian Shaw, *Predator Empire: Drone Warfare and Full Spectrum Dominance* (Minneapolis: University of Minnesota Press, 2016); Derek Gregory, "From a View to a Kill: Drones and Late Modern War," *Theory, Culture & Society* 28, nos. 7–8 (2014): 193; Katharine Hall Kindervater, "The Emergence of Lethal Surveillance: Watching and Killing in the History of Drone Technology," *Security Dialogue* 47, no. 3 (2016): 223–238; Benjamin Noys, "Drone Metaphysics," *Culture Machine* 16 (2015): 1–22; Eyal Weizman, *Forensic Architecture: Violence at the Threshold of Detectability* (New York: Zone, 2019); Lisa Parks, "Drones, Vertical Mediation, and the Targeted Class," *Feminist Studies* 42, no. 1 (2016): 227–35; Daniel Greene, "Drone Vision," *Surveillance & Society* 13, no. 2 (2015): 233–249; Stephen Graham, "Drone: The Robot Imperium," in *Vertical: The City from Satellites to Bunkers* (London: Verso, 2016), 67–94.

3. This term was coined by Christian Metz, whose work is germinal in this regard. Christian Metz, *Film Language: A Semiotics of the Cinema* (Chicago: University of Chicago Press, 1991), first published 1971 in French.

4. For more on the notion of the scopic regime, see Martin Jay, "The Scopic Regimes of Modernity," in *Vision and Visuality*, ed. Hal Foster (Seattle: Bay Press, 1988), 3–29. Jay has also suggested that an unconditional culturalization and relativization of vision can lead to cultural universalism: Martin Jay, "Cultural Relativism and the Visual Turn," *Journal of Visual Culture* 1, no. 3 (2002): 267–278. For further discussion and critique of the scopic regime and the historicization of vision, see Jonathan Crary, *Techniques of the Observer: On Vision and Modernity* (Cambridge, MA: MIT Press, 1992); Antonio Somaini, "On the Scopic Regime," *Leitmotiv* 5 (2005–2006): 25–38.

5. As one example of this connection, see Desmond Manderson, "Chronotopes in the Scopic Regime of Sovereignty," *Visual Studies* 32, no. 2 (2017): 167–177.

6. I owe this insight to Antoine Bousquet's *The Eye of War: Military Perception from the Telescope to the Drone* (Minneapolis: University of Minnesota Press, 2018), 10.

7. Paul Virilio, *War and Cinema: The Logistics of Perception* (London: Verso, 1989). See also Anders Engberg-Pedersen and Kathrin Maurer, eds., *Visualizing War: Images, Emotions, Communities* (New York: Routledge, 2018).

8. For a discussion of this, see Caren Kaplan, "Mobility and War: The Cosmic View of US Air Power," *Environment and Planning* 38, no. 2 (2006): 395–407.

9. Caren Kaplan, *Aerial Aftermaths: Wartime from Above* (Durham, NC: Duke University Press, 2018), 9.

10. Chamayou, *Drone Theory*; Holland, *Eyes in the Sky*; Adey, Whitehead, and Williams, "Introduction"; Hugh Gusterson, *Drone*; Shaw, *Predator Empire*; Kindervater, "Emergence of Lethal Surveillance"; Noys, "Drone Metaphysics"; Weizman, *Forensic Architecture*; Parks, "Drones."

11. Gregory, "From a View," 193.

12. E.g., Lisa Parks, "Vertical Mediation and the U.S. Drone War in the Horn of Africa," in *Life in the Age of Drone Warfare*, ed. Caren Kaplan and Lisa Parks (Durham, NC: Duke University Press, 2017), 134–158; Bousquet, *The Eye of War*; Kathrin Maurer, "Visual Power: The Scopic Regime of Military Drones," *War, Media, Conflict* 10, no. 2 (2017): 141–151. Steyerl articulated a critique of the vertical paradigm of the military drone: Hito Steyerl, "In Free Fall: A Thought Experiment on Vertical Perspective," in *To See without Being Seen: Contemporary Art and Drone Warfare*, ed. Svea Braeunert and Meredith Malone (Chicago: University of Chicago Press, 2016), 71–81; Hito Steyerl "In Free Fall: A Thought Experiment on Vertical Perspective," *e-flux journal* 24 (2011): 1–11. See also Thomas Stubblefield, *Drone Art: The Everywhere War as Medium* (Oakland: University of California Press, 2020). See also Sarah Tuck, ed., *Drone Vision: Warfare, Surveillance, Protest* (Gothenburg: Art and Theory Publishing Hasselblad Foundation, 2022).

13. Ciara Bracken-Roche, "Domestic Drones: The Politics of Verticality and the Surveillance Industrial Complex," *Geographica Helvetica* 71, no. 3 (2016): 167–172.

14. Weizman, *Forensic Architecture*.

15. Roger Stahl, "What the Drone Saw: The Cultural Optics of the Unmanned War," *Australian Journal of International Affairs* 67, no. 5 (2013): 668. See also Green, "Drone Vision"; Joanne McNeil and Ingrid Burrington, "Dronism," *Dissent* 61, no. 2 (2014): 57–60.

16. Ludwig Wittgenstein, *Philosophische Untersuchungen* (Frankfurt: Suhrkamp, 2001), 786–787. I owe this idea of family resemblances between civilian and military drones to Maximilian Jablonowski, "Dronie Citizenship?" in *Selfie Citizenship*, ed. Adi Kuntsman (London: Palgrave Macmillan, 2017), 97–106.

17. See Friedrich Kittler, *Gramophone, Film, Typewriter* (Stanford: Stanford University Press, 1999).

18. See e.g., Max Liljefors, who criticizes the fetishization of the military drone as an all-seeing technology: Max Liljefors, "Omnivoyance and Blindness," in *War and Algorithm*, ed. Max Liljefors, Gregor Noll, and Daniel Steuer (London: Rowman and Littlefield, 2020), 127–164. Although my book focuses on civilian drones, I would also expand my argument to suggest that military drones themselves do not exclusively see and execute violence according to the scopic regime of verticality.

19. My critique of the aerial is inspired by Somaini's "On the Scopic Regime" and Kaplan's *Aerial Aftermaths*. The art project *Decolonized Sky 2014*, curated by Gilead Reich and Yael Messer, likewise aims to show that artists and activists in recent years have reworked aerial perspectives of power and control to develop a new language of ethics and aesthetics. "Decolonized Skies," Gilad Reich, accessed October 27, 2021, https://giladreich.net/Decolonized-Skies.

20. Denis Cosgrove and Carmen Cosgrove, *Apollo's Eye: A Cartographic Genealogy of the Earth in the Western Imagination* (Baltimore: Johns Hopkins University Press, 2003), xi.

21. Augustine, *The Confessions of St. Augustine*, trans. Rex Warner (New York: Penguin Putnam, 2001), 165.

22. Michael Valaouris, *Das Feld hat Augen: Bilder des überwachenden Blicks* (Berlin: Deutscher Kunstverlag Kunstbibliothek Staatliche Museen zu Berlin, 2017), 13–14.

23. For a discussion of Descartes's experiment, see Claus Zittel, *Theatrum Philosophicum: Descartes und die Rolle ästhetischer Formen in der Wissenschaft* (Berlin: Akademie, 2009).

24. Somaini, "On the Scopic Regime," 36.

25. As Jay notes, "Cartesian perspectivalism in fact may nicely serve as a shorthand way to characterize the dominant scopic regime of the modern era." Martin Jay, *Downcast Eyes: The Denigration of Vision in Twentieth-Century French Thought* (Berkeley: University of California Press, 1993), 69–70.

26. Valaouris, *Das Feld hat Augen*, 17–18.

27. Michel Foucault, *Discipline and Punish: The Birth of the Prison* (New York: Pantheon, 1977).

28. See also the history of modern vision and hegemony in Michael Levin, *Modernity and the Hegemony of Vision* (Berkeley: University of California Press, 1993), 212.

29. Adey, Whitehead, and Williams, "Introduction," 176.

30. This connection is also emphasized in Astrit Schmidt-Burckhardt's "The All-Seer: God's Eye as Proto-Surveillance," in *CRTL [SPACE]: Rhetorics of Surveillance from Bentham to Big Brother*, ed. Thomas Y. Levin, Ursula Frohne, and Peter Weibel (Cambridge, MA: MIT Press, 2002), 16–31.

31. See also Stubblefield, *Drone Art*, 103–133.

32. Joanna Zylinska, *Nonhuman Photography* (Cambridge, MA: MIT Press, 2017), 16; Donna Haraway, "Situated Knowledges: The Science Question in Feminism and the Privilege of Partial Perspective," *Feminist Studies* 14, no. 3 (1988): 575–599.

33. Zylinska, *Nonhuman Photography*, 16.

34. Zylinska, *Nonhuman Photography*, 15.

35. For a discussion of the drone as an unstable technology, see Jordan Crandall, "Ecologies of a Wayward Drone," in *From Above: War, Violence and Verticality*, ed. Peter Adey, Mark Whitehead, and Alison J. Williams (London: Hurst, 2011), 263–287.

36. Bradley L. Garret and Anthony McCosker connect this idea of the assemblage to the drone: Bradley L. Garret and Anthony McCosker, "Non-human Sensing: New Methodologies for the Drone Assemblage," in *Refiguring Techniques in Digital Visual Research*, ed. Edgar Gómez Cruz, Shanti Sumartojo, and Sarah Pink (London: Palgrave Macmillan, 2017), 16. Jeremy Crampton also refers to the drone assemblage, although given the drone's nonscopic and multisensorial configurations, I would not follow him in terming it a *vertical* assemblage: Jeremy Crampton, "Assemblage of the Vertical: Commercial Drones and Algorithmic Life," *Geographica Helvetica* 71 (2016): 137–146. For a more general idea of research on sensing, I have been guided by David Howes and Constance Classen, *Ways of Sensing: Understanding the Senses in Society* (London: Routledge, 2013).

37. Abhishek Anand et al., "Protocol Development for Real-Time Ship Fuel Sulfur Content Determination Using Drone-Based Plume Sniffing Microsensor System," *Science of the Total Environment* 744 (2020): 140885.

38. Lisa Parks, "Drones, Infrared Imagery, and Body Heat," *International Journal of Communication* 8 (2014): 2518–2521.

39. Anna Jackman, "Sensing," Society for Cultural Anthropology, *Theorizing the Contemporary, Fieldsights* (blog), June 27, 2017, https://culanth.org/fieldsights/sensing;

Jan Mieszkowski, "The Drone of Data," in *Drone Imaginaries: The Remote Power of Vision*, ed. Andreas Immanuel Graae and Kathrin Maurer (Manchester: Manchester University Press, 2021), 55–73.

40. Maximilian Jablonowski, "Ferngesteuertes Feeling: Zur Technogenen Sensualität Unbemannten Fliegens," in *Kulturen der Sinne: Zugänge zur Sensualität der Sozialen Welt*, ed. Karl Braun et al. (Würzburg: Königshausen & Neumann), 385.

41. Jackman, "Sensing."

42. Shaun Gallagher and Jonathan Cole discuss the kinetic aspect of embodiment and its relation to empathy: Shaun Gallagher and Jonathan Cole, "Body Schema and Body Image in a Deafferented Subject," *Journal of Mind and Behavior* 16, no. 4 (1995): 369–390.

43. Walter Benjamin, "The Work of Art in the Age of Mechanical Reproduction," in *Illuminations: Essays and Reflections*, ed. Hannah Arendt, trans. Harry Zohn (New York: Schocken, 1968), 39.

44. Crary, *Techniques of the Observer*.

45. Michael Bull et al., "Introducing Sensory Studies," *The Senses and Society* 1, no. 1 (2006): 5.

46. On this intertwinement of humans and drones, and on the dangers of seeing drones as strictly nonhuman, see Nina Franz and Moritz Queisner, "Die Akteure verlassen die Kontrollstation: Krisenhafte Kooperationen im bildgeführten Drohnenkrieg," in *Das Mitsein der Medien: Prekäre Koexistenzen von Menschen, Maschinen und Algorithmen*, ed. Johannes Bennke et al. (Munich: Fink, 2018), 27–58.

47. Franz and Queisner, "Die Akteure," 32. For an analysis of the role of human personnel and their labor in military ground stations, see Peter Asaro, "The Labor of Surveillance and Bureaucratized Killing: New Subjectivities of Military Drone Operators," *Social Semiotics* 23, no. 2 (2013): 1–29.

48. I owe this insight to Jablonowski, "Ferngesteuertes Feeling," 386.

49. Gilles Deleuze and Félix Guattari, *A Thousand Plateaus: Capitalism and Schizophrenia* (Minneapolis: University of Minnesota Press, 1987), first published 1980 in French.

50. In other words, human agents and machine technology form what Lucy Suchman has called a *configuration*, in which both can change, adapt, and take on different positions: Lucy Suchman, *Human-Machine Reconfigurations: Plans and Situated Actions* (New York: Cambridge University Press, 2007).

51. Gilbert Simondon, *On the Mode of Existence of Technical Objects*, trans. Cecile Malaspina and John Rogove (Minneapolis: University of Minnesota Press, 2017), 15.

52. Martin Heidegger, "Die Frage nach der Technik, Vorträge und Aufsätze (1910–1976)," in *Gesamtausgabe*, vol. 7, ed. Friedrich Wilhelm von Herrmann (Frankfurt: Vittorio Klostermann, 2000), 5–36.

53. Manfred Faßler, "Stile der Anwesenheit: Technologien, Traumgesichter, Medien," in *Wunschmaschine/Welterfindung: Eine Geschichte der Technikvisionen seit dem 18. Jahrhundert*, ed. Brigitte Felderer (Vienna: Springer, 1996), 251–271.

54. Lucy Suchman and Jutta Weber debate notions of autonomy in relation to remote technology, tracing autonomy as an Enlightenment concept: Lucy Suchman and Jutta Weber, "Human-Machine Autonomies," in *Autonomous Weapon Systems: Law, Ethics, Policy*, ed. Bhuta Nehal et al. (New York: Cambridge University Press, 2016), 75–102.

55. Jens Eder and Charlotte Klonk, ed., *Image Operations: Visual Media and Political Conflict* (Manchester: Manchester Press, 2016). This anthology provides great insights into discussions of remote images in military, aesthetic, and medical contexts.

56. Harun Farocki, "Phantom Images," trans. Brian Poole, *Public* 29 (2004): 12–24.

57. On digital images and machine vision as representation, see Ingrid Hoelzl, "Postimage," in *Posthuman Glossary*, ed. Rosi Braidotti and Maria Hlavajova (London: Bloomsbury, 2018), 360–362.

58. On military drones as data archives, see Kathrin Maurer, "Drones as Big Data Archives: Mimesis and Counter-Archiving in Contemporary Art on Military Drones," in *(W)archives*, ed. Daniela Agostinho et al. (Berlin: Sternberg Press, 2020), 119–141.

59. For an extended discussion of images and operations in the context of new media, see Lev Manovich, *The Language of New Media* (Cambridge, MA: MIT Press, 2005).

60. Gregory, "From a View," 192.

61. Lisa Gitelman, ed., *Raw Data Is an Oxymoron* (Cambridge, MA: MIT Press, 2013).

62. Donna Haraway, *Manifestly Haraway* (Minneapolis: University of Minnesota Press, 2016), 3.

63. On cyborg research in cultural studies, gaming theory, medicine, anthropology, media, and science, see Niall Richardson and Adam Locks, *Body Studies: The Basics* (New York: Routledge, 2014), 95; Gustav Roßler, "Kleine Galerie neuer Dingbegriffe: Hybriden, Quasi-Objekte, Grenzobjekte, epistemische Dinge," in *Bruno Latours Kollektive: Kontroversen zur Entgrenzung des Sozialen*, ed. Georg Kneer, Markus Schroer, and Erhard Schüttpelz (Frankfurt: Suhrkamp, 2008), 76–107.

64. Haraway, "Situated Knowledges," 581.

65. Haraway, "Situated Knowledges," 582.

66. Haraway, "Situated Knowledges," 582.

67. "The moral is simple. Only partial perspective promises objective vision." Haraway, "Situated Knowledges," 583.

68. See also Katherine Chandler, "A Drone Manifesto: Re-forming the Partial Politics of Targeted Killing," *Catalyst: Feminism, Theory, and Technoscience* 2, no. 1 (2016):

1–23; Anna Feigenbaum, "From Cyborg Feminism to Drone Feminism: Remembering Women's Anti-Nuclear Activisms," *Feminist Theory* 16, no. 3 (2015): 265–288.

69. Alexander R. Galloway and Eugene Thacker, *The Exploit: A Theory of Networks* (Minneapolis: University of Minnesota Press, 2013), 5. See also Thomas Stubblefield, "In Pursuit of Other Networks," in Caplan and Parks, *Life in the Age*, 195–219.

70. Chandler, "A Drone Manifesto"; Stubblefield, *Drone Art*; Caroline Holmqvist, "Undoing War: War Ontologies and the Materiality of Drone Warfare," *Millennium* 41, no. 3 (2014): 535–552. These works are particularly important to my approach because they also critique the military drone's scopic regime by pointing to its networkcentric configuration.

71. Zylinska, *Nonhuman Photography*, 8.

72. Zylinska, *Nonhuman Photography*, 23

73. The sensible encompasses all that is there to sense, all that is sayable and visible within society: "The distribution of the sensible is the system of self-evident facts of sense perception that simultaneously discloses the existence of something in common and the delimitations that define the respective parts and positions within it." Jacques Rancière, *The Politics of Aesthetics*, trans. Gabriel Rockhill (London: Bloomsbury, 2010), 12.

74. Jacques Rancière, *Dissensus: On Politics and Aesthetics*, ed. and trans. Steven Corcoran (London: Continuum, 2010), 7.

75. Rancière, *Dissensus*, 176

76. Rancière's ideas about the sensible and aesthetics can be compared to Kant's notion of the *sensus communis* as a faculty of aesthetic judgment. According to Kant, aesthetic judgments are manifested by the sharing of a common taste, opinion, or beautiful experience, and art can evoke and harbor this *sensus communis*. In contrast to Kant, however, Rancière does not look for consensus; rather, the experience of something in common opens up a trajectory for critique of the common idea of the sensible. See also John D. Schaefer, *Sensus Communis: Vico, Rhetoric, and the Limits of Relativism* (Durham, NC: Duke University Press, 1990); Lauren Berlant, "The Commons: Infrastructures for Troubling Times," *Environment and Planning: Society and Space* 34, no. 3 (2016): 393–419.

77. Ferdinand Tönnies, "Gemeinschaft und Gesellschaft (Erstausgabe 1887)," in *Ferdinand Tönnies Gesamtausgabe*, vol. 2, ed. Bettina Clausen and Dieter Haselbach (Berlin: Walter de Gruyter, 2019), 3–34.

78. Jean-Luc Nancy, *The Inoperative Community* (Minneapolis: University of Minnesota Press, 1991); Jean-Luc Nancy, "The Confronted Community," *Postcolonial Studies* 6, no. 1 (2003): 23–36. It is important to note that Nancy's ideas about community are embedded in the work of a whole raft of thinkers, including Jacques Derrida, Georges Bataille, Maurice Blanchot, and Giorgio Agamben.

79. Nancy, *Inoperative Community*, 15.

80. "What this community has lost—the immanence and the intimacy of a communion—is lost only in the sense that such a 'loss' is constitutive of 'community' itself." Nancy, *Inoperative Community*, 12.

81. Storytelling as the *Urszene* of mythmaking can also undo and interrupt myth, a process that in turn opens up a constructive moment for Nancy's idea of community. In storytelling's performative setting, the listeners—or as Nancy would say, the singularities—want to hear about the presence of the myth, but they can only experience its absence. They are temporally far away from the age of myth, and they experience their own finitude and mortality in contrast to myth's eternal nature: while listening to the myth, they are distanced both from the story and from themselves simultaneously. In this fragile and fluid moment—a hermeneutic moment of worldmaking closely related to the Heideggerian view of understanding—we interrupt the myth, but we also shape a provisional, temporary, flowlike, hovering state of being-together, which in turn renegotiates, touches, limits, and improvises our relationships with each other.

82. In addition to the communities found in aesthetic fictions, I argue that the communities shaped by people who fly, operate, and use drones also have an imaginary element. As I show in chapter 2, hobby dronists and drone racers share a vision of an imagined community together with their drones. The police drones that are used to surveil neighborhoods via facial recognition software, and which are supposed to predict crime, are also based on visions of imaginary communities (see chapter 3).

83. Rosi Braidotti, *The Posthuman* (Cambridge, UK: Polity, 2013).

84. Rosi Braidotti, *Posthuman Knowledge* (Cambridge, UK: Polity, 2019), 3.

85. Rosi Braidotti, *Metamorphoses: Towards a Materialist Theory of Becoming* (Hoboken: John Wiley & Sons, 2013), 3.

86. Braidotti, *Posthuman Knowledge*, 47.

87. Braidotti, *Posthuman Knowledge*, 54.

88. Braidotti, *Posthuman Knowledge*, 49.

89. Braidotti, *Posthuman Knowledge*, 52.

90. Mark Andrejevic describes "the drone as an avatar of the emerging logic of increasingly passive interactivity." Mark Andrejevic, "Theorizing Drones and Droning Theory," in *Drones and Unmanned Aerial Systems: Legal and Social Implications for Security and Surveillance*, ed. Aleš Završnik (Cham: Springer, 2016), 21.

91. Andrejevic, "Theorizing Drones," 21.

92. On data capitalism, see Shosana Zuboff, *The Age of Surveillance Capitalism: The Fight for a Human Future at the New Frontier of Power* (London: Profile, 2019). In later

chapters, I discuss aesthetic drone imaginaries that format inhuman communities according to the zoological formation of the swarm (chapter 6) as well as facial recognition and algorithms (chapter 3).

Chapter 2

1. Don Ihde, *Bodies in Technology* (Minneapolis: University of Minnesota Press, 2002).

2. I am indebted to Maximilian Jablonowski, whose research (and our discussions of his work) gave me an insight into the relationship between drones and telepresence.

3. Oliver Grau, *Virtual Art: From Illusion to Immersion* (Cambridge, MA: MIT Press, 2003), 278.

4. Grau, *Virtual Art*, 279.

5. Grau refers to Hans Belting's anthropological approach to images. Grau, *Virtual Art*, 279.

6. See my research on this: Kathrin Maurer, *Visualizing the Past: The Power of the Image in German Historicism* (Berlin: Walter de Gruyter, 2013).

7. Christian Bailly, *Automata: The Golden Age, 1848–1914* (London: Robert Hale, 2003).

8. Grau, *Virtual Art*, 285.

9. *Merriam-Webster's Collegiate Dictionary*, 11th ed. (Springfield, MA: Merriam-Webster, 2003), 895.

10. On the history of the panorama in Germany, see Stephan Oettermann, *The Panorama: History of a Mass Medium*, trans. Deborah Lucas-Schneider (New York: Zone, 1997). See also Heinz Buddemeier, *Panorama, Diorama, Photographie: Entstehung und Wirkung neuer Medien im 19. Jahrhundert* (Munich: Fink, 1977); Ralph Hyde, *Panoramania! The Art and Entertainment of the All-Embracing View* (London: Trefoil, 1988).

11. I have written about the panorama as an immersive medium but not related it to the drone. See Kathrin Maurer, "The Paradox of Total Immersion: Watching War in Nineteenth-Century Panoramas," in *Visualizing War: Images, Emotions, Communities*, ed. Anders Engberg-Pedersen and Kathrin Maurer (New York: Routledge, 2018), 78–94.

12. Jonathan Crary, *Techniques of the Observer: On Vision and Modernity* (Cambridge, MA: MIT Press, 1992), 113.

13. This roving gaze is structurally similar to the filmic eye incorporated into the gaze through a train window. Wolfgang Schivelbusch makes this connection between train travel and panoramic paintings in his book *The Railway Journey* (New York: Urizen, 1980).

14. Oettermann, *The Panorama*, 41.

15. Maximilian Jablonowski, "Beyond Drone Vision: The Embodied Telepresence of First-Person-View Drone Flight," *The Senses and Society* 15, no. 3 (2020): 344–358.

16. Airvuz, accessed June 9, 2022, https://www.airvuz.com/.

17. I owe much of my knowledge about hobby drone cultures to Jablonowski's work: Maximilian Jablonowski, "Drone It Yourself! On the Decentering of Drone Stories," *Culture Machine* 16 (2015), http://culturemachine.net/drone-culture/drone-ityourself/.

18. On the cultural context of levitators and drones, see Peter Adey, "Making the Drone Strange: The Politics, Aesthetics and Surrealism of Levitation," *Geographica Helvetica* 71, no. 4 (2016): 319–329.

19. "Our ideas of levitators come from a wide range of traditions in philosophy, theology, politics, science and visual culture. Christian, and almost every other major faith's, narratives of ascension would follow the rising of the soul to join with God, commonly allegorized in pilgrimages, rituals and rites of passage up hills, steps and mountains and, ultimately, floating figures." Adey, "Making the Drone Strange," 320.

20. On the colonialist optics of the panorama, see Linda Nochlin, *The Politics of Vision: Essays on Nineteenth-Century Art and Society* (Boulder: Westview Press, 1989).

21. On the drone's dependency on its environment, see Anthony McCosker, "Drone Media: Unruly Systems, Radical Empiricism, and Camera Consciousness," *Culture Machine* 16 (2015), https://culturemachine.net/drone-culture/drone-media/.

22. For a discussion of the bodily effects of drone racing, see Jablonowski, "Drone It Yourself!"

23. Steen Ledet Christiansen, "Unruly Vision: Synesthetic Space: Drone Music Videos," *The Senses and Society* 15, no. 3 (2020): 286–298.

24. Lindsay Jacobson, "Be One with the Drone: Elite Athletes Compete Head-to-Head in Drone Racing League," *ABC News*, July 27, 2019, https://abcnews.go.com/Technology/drone-elite-athletes-compete-head-head-drone-racing/story?id=64494623.

25. Heather Millar, "Racing Drones," *Air & Space Magazine*, November 2016, https://www.airspacemag.com/flight-today/racing-drones-180960969/.

26. FPV Drone Pilots forum, accessed June 9, 2022, https://fpvdronepilots.com/.

27. Michael Salter, "Toys for the Boys? Drones, Pleasure and Popular Culture in the Militarisation of Policing," *Critical Criminology* 22, no. 2 (2014): 163–177.

28. The Drone Girl, accessed June 30, 2020, http://thedronegirl.com/.

29. Sally French, "Meet the World's Top Female Drone Pilot," The Drone Girl, April 25, 2016, http://thedronegirl.com/2016/04/25/meet-zoe-the-worlds-top-female-drone-pilot/.

30. According to Ehemann, they can be productive in therapy, reflecting the self from new perspectives. Rose Ehemann, "Selfies and Dronies: Zu Chancen und Gefährdungen der Dynamisierung des Selbst in virtuellen Welten," *Journal für Psychoanalyse* 59 (2018): 60–73.

31. Jacques Lacan, "The Subversion of the Subject and the Dialectic of Desire in the Freudian Unconscious," in *Écrits: A Selection* (New York: Routledge, 1977), 323–360.

32. Ehemann, "Selfies and Dronies," 67–68.

33. I owe this reference to Sloterdijk to Ehemann, "Selfies and Dronies," 62, and to Jablonowski, "Dronie Citizenship?" 99.

34. See, e.g., Hanna Brooks Olsen, "Dronie Like a Pro: How to Master the Drone Selfie," CreativeLive, July 21, 2014, https://www.creativelive.com/blog/dronie-tips.

35. Although, as David Beesley has shown, there are politized drone race communities. David Beesley, "Drone Panic! On Representations of the Personal Drone by Australian Mainstream Media," *Screen Thought* 1, no. 1 (2018): 1–18.

36. E.g., the artists Eduardo Kac and Ken Goldberg were central in this movement, as their works engaged with immersive environments and telecommunication. An important forerunner was Roy Ascott's telematic artwork *La Plissure du Texte* (1983), an installation consisting of sixteen speakers all over the world that were connected to each other by telephone, with each speaker assigned its own role. See Grau, *Virtual Art*, 270–94.

37. Jablonowski, "Ferngesteuertes Feeling."

38. Juliane Rebentisch, *Ästhetik der Installation* (Frankfurt: Suhrkamp, 2003).

39. Amely Deiss et al., eds., *Raphaela Vogel* (Cologne: Walther König, 2018).

40. Bernd Lechler, "Raphaela Vogel, Drohnen: Ungeheuer am Himmel?" *Deutschlandfunk*, June 6, 2019, https://www.deutschlandfunk.de/raphaela-vogel-bei-game-of-drones-drohnen-ungeheuer-am.807.de.html?dram:article_id=450868, my translation.

41. See the discussion of *Seagulls* in Thomas Stubblefield, *Drone Art: The Everywhere War as Medium* (Oakland: University of California Press, 2020), 112–116, 121–123.

42. Haus der Kunst, *Künstlergespräch mit Raphaela Vogel und der Philosophin Juliane Rebentisch*, YouTube, 1:14:03, uploaded July 1, 2019, https://www.youtube.com/watch?v=7bkiHozNxOk, my translation.

43. Queering in the context of military drones has been discussed by Cara Daggett and Ronak K. Kapadia, who show that drone warfare can upset traditional visions of masculinity. Ronak K. Kapadia, *Insurgent Aesthetics: Security and the Queer Life of the Forever War* (Durham, NC: Duke University Press, 2019); Cara Daggett, "Drone

Disorientations: How 'Unmanned' Weapons Queer the Experience of Killing in War," *International Feminist Journal of Politics* 17, no. 3 (2015): 361–379.

44. "Korakrit Arunanondchai: Painting with History in a Room Filled with People with Funny Names 3 at Palais de Tokyo Paris," *Mousse Magazine*, 2015, http://moussemagazine.it/korakrit-arunanondchai-palaisdetokyo.

45. Thom Bettridge, "The Fabric of Existence: A Studio Visit with Denim Painter Korakrit Arunanondchai in New York," Ssense.com, accessed October 2, 2021, https://www.ssense.com/en-us/editorial/fashion/the-fabric-of-existence.

46. Animism is discussed by Grant Bollmer and Katherine Guinness, "Do You Really Want to Live Forever? Animism, Death, and the Trouble of Digital Images," *Cultural Studies Review* 24, no. 2 (2018): 79–96.

47. Bettridge, "The Fabric of Existence."

48. Bettridge, "The Fabric of Existence."

49. Martha Kirszenbaum, "Korakrit Arunanondchai by Martha Kirszenbaum," *Bomb* 149 (2019), https://bombmagazine.org/articles/korakrit-arunanondchai/Kirszenbaum.

50. Jablonowski, "Beyond Drone Vision," 354.

51. Roy Ascott, "Telenoia," in *Telematic Embrace: Visionary Theories of Art, Technology, and Consciousness*, ed. Edward A. Shanken (Berkeley: University of California Press, 2003), 266.

52. Marvin Minsky, "Telepresence," *Omni Magazine*, June 1980, https://web.media.mit.edu/~minsky/papers/Telepresence.html.

53. Kana Misawa and Jun Rekimoto, "ChameleonMask: Embodied Physical and Social Telepresence Using Human Surrogates," in *Proceedings of the 33rd Annual ACM Conference: Extended Abstracts on Human Factors in Computing Systems*, ed. Bo Begole and Jinwoo Kim (New York: Association for Computing Machinery (2015), 401–411.

54. "ChameleonMask: Embodied Physical and Social Telepresence Using Human Surrogates," Rekimoto Lab, 2015, https://lab.rekimoto.org/projects/chameleonmask/.

55. For a theoretical discussion of the notion of exploitation, see Jon Elster, "Exploring Exploitation," *Journal of Peace Research* 15, no. 1 (1978): 3–17.

56. Donna Haraway, "Situated Knowledges: The Science Question in Feminism and the Privilege of Partial Perspective," *Feminist Studies* 14, no. 3 (1988): 582.

57. *Sleep Dealer*, directed by Alex Rivera, Maya Entertainment, 2008.

58. Altha Cravey, Joseph Palis, and Gabriela Valdivia, "Imagining the Future from the Margins: Cyborg Labor in Alex Rivera's *Sleep Dealer*," *GeoJournal* 80 (2015): 867–880; Libia Villazana, "Transnational Virtual Mobility as a Reification of Deployment

of Power: Exploring Transnational Processes in the Film *Sleep Dealer*," *Transnational Cinemas* 4, no. 2 (2013): 217–230.

59. Rosi Braidotti, *Metamorphoses: Towards a Materialist Theory of Becoming* (Hoboken: John Wiley & Sons, 2013), 226.

60. Donna Haraway, *Manifestly Haraway* (Minneapolis: University of Minnesota Press, 2016), 28.

Chapter 3

1. Nicholas Mirzoeff, "The Right to Look," *Critical Inquiry* 37 no. 3 (2011): 473–496. The impossibility of looking back constitutes a key scopic regime of colonialism and imperialism and is also reflected in the visual constellations of drone warfare. See also Kathrin Maurer, "Visual Power: The Scopic Regime of Military Drones," *War, Media, Conflict* 10, no. 2 (2017): 141–151; Derek Gregory, "The Everywhere War," *Geographical Journal* 177, no. 3 (2011): 238–250.

2. Emanuel Levinas, *Ethics and Infinity: Conversations with Philippe Nemo* (Pittsburgh: Duquesne University Press, 2011), 89. I owe this reference to Thomas Stubblefield, *Drone Art: The Everywhere War as Medium* (Oakland: University of California Press, 2020), 124–125.

3. Levinas, *Ethics and Infinity*, 89.

4. This ethical debate about drones is particularly evident in the work of two opposing scholars. Grégoire Chamayou argues that drone warfare engages in "necroethics" and "manhunting"; Bradley Strawser states by contrast that drone warfare is ethical and that one even has a moral obligation to engage in remote warfare. Grégoire Chamayou, *A Theory of the Drone* (New York: New Press, 2015); Bradley J. Strawser, "Moral Predators: The Duty to Employ Uninhabited Aerial Vehicles," *Journal of Military Ethics* 9, no. 4 (2015): 342–368.

5. Matt Delmont, "Drone Encounters: Noor Behram, Omer Fast, and Visual Critiques of Drone Warfare," *American Quarterly* 65, no. 1 (2013): 193–202. Stubblefield reads this picture differently, as revealing the networkcentric quality of modern warfare: Stubblefield, *Drone Art*, 30–31.

6. Samuel Gibbs, "Google's AI Is Being Used by US Military Drone Programme," *The Guardian*, March 7, 2018, https://www.theguardian.com/technology/2018/mar/07/google-ai-us-department-of-defense-military-drone-project-maven-tensorflow.

7. Adam Frisk, "What Is Project Maven? The Pentagon AI Project Google Employees Want Out Of," *Global News*, April 5, 2018, https://globalnews.ca/news/4125382/google-pentagon-ai-project-maven/; Zachary Fryer-Biggs, "Inside the Pentagon's Plan to Win over Silicon Valley's AI Experts," *Wired*, December 21, 2018, https://www.wired.com/story/inside-the-pentagons-plan-to-win-over-silicon-valleys-ai-experts/.

8. Scott Shane and Daisuke Wakabayashi, "The Business of War: Google Employees Protest Work for the Pentagon," *New York Times*, April 4, 2018, https://www.nytimes .com/2018/04/04/technology/google-letter-ceo-pentagon-project.html.

9. "7 Best Drones with Facial Recognition," The Droid Guy, May 2022, https:// thedroidguy.com/7-best-drones-with-facial-recognition-1077684.

10. John Paulin Hansen et al., "The Use of Gaze to Control Drones," in *Proceedings of the Symposium on Eye Tracking Research and Applications*, ed. Pernilla Qvarfordt and Dan Wintzer Hansen (New York: The Association of Computing Machinery, 2014), 27–34.

11. Hwai-Jung Hsu and Kuan-Ta Chen, "Face Recognition on Drones: Issues and Limitations," in *Proceedings of the First Workshop on Micro Aerial Vehicle Networks, Systems, and Applications for Civilian Use*, ed. Kuan-Ta Chen et al. (New York: The Association for Computing Machinery, 2015), 39–44.

12. Faine Greenwood, "Can a Police Drone Recognize Your Face?" *Slate*, July 8, 2020, https://slate.com/technology/2020/07/police-drone-facial-recognition.html.

13. Greenwood, "Can a Police Drone?"

14. Greenwood, "Can a Police Drone?"

15. Greenwood, "Can a Police Drone?"

16. James L. Wayman, "The Scientific Development of Biometrics over the Last 40 Years," in *The History of Information Security*, ed. Karl De Leeuw and Jan Bergstra (Amsterdam: Elsevier Science, 2007), 263–274. For further reading on biometrics, see Paul Reid, *Biometrics for Network Security* (Upper Saddle River: Prentice Hall, 2004).

17. "Strictly speaking, facial recognition technology treats the face as an index of identity, disregarding its expressive capacity and communicative role in social interaction." Kelly A. Gates, *Our Biometric Future: Facial Recognition Technology and the Culture of Surveillance* (New York: New York University Press, 2011), 8.

18. Gates, *Our Biometric Future*, 10.

19. Olivia Varley-Winter, "The Overlooked Governance Issues Raised by Facial Recognition," *Biometric Technology Today* 5 (2020): 5–8. Similarly controversial are the facial recognition companies Face++ and Amazon Rekognition.

20. See Roberto Brunelli and Tomaso Poggio, "Face Recognition: Features versus Templates," *IEEE Transactions on Pattern Analysis and Machine Intelligence* 15, no. 10 (1993): 1042–1052.

21. Gates, *Our Biometric Future*, 30.

22. Lila Lee-Morrison, *Portraits of Automated Facial Recognition: On Machinic Ways of Seeing the Face* (Bielefeld: Transcript, 2019).

23. Lee-Morrison, *Portraits*, 55–84.

24. Peter N. Belhumeur, Joao P. Hespanha, and David J. Kriegman, "Eigenfaces vs. Fisherfaces: Recognition Using Class Specific Linear Projection," *IEEE Transactions on Pattern Analysis and Machine Intelligence* 19, no. 7 (1997): 711–720.

25. Tom Simonite, "Facebook Creates Software That Matches Faces Almost as Well as You Do," *Technology Review*, March 17, 2014, https://www.technologyreview.com/2014 /03/17/13822/facebook-creates-software-that-matches-faces-almost-as-well-as-you-do.

26. See Katrin Amelang, "Zur Sinnlichkeit von Algorithmen und ihrer Erforsch-barkeit," in *Kulturen der Sinne: Zugänge zur Sensualität der sozialen Welt*, ed. Karl Braun et al. (Würzburg: Königshausen Neumann, 2017), 358–367. In a military context, see Louise Amoore, "Algorithmic War: Everyday Geographies of the War on Terror," *Antipode* 41, no. 1 (2009): 49–69.

27. Hermann von Helmholtz, *Die Thatsachen in der Wahrnehmung* (Berlin: August Hirschwald, 1879), 12, my translation.

28. Trevor Paglen, "Invisible Images: Your Pictures Are Looking at You," *Architectural Design* 89, no. 1 (2019): 27.

29. Lee-Morrison, *Portraits*, 159–177.

30. Paglen, "Invisible Images," 26

31. Ingrid Hoelzl and Remi Marie, "From Softimage to Postimage," *Leonardo* 50, no. 1 (2017): 72–73. For an overview and discussion of machinic images, see Aud Sissel Hoel, "Operative Images: Inroads to a New Paradigm of Media Theory," in *Image-Action-Space: Situating the Screen in Visual Practice*, ed. Louisa Feiersinger, Kathrin Friedrich, and Moritz Queisner (Berlin: Walter de Gruyter, 2018), 11–28.

32. This idea of images as performative media that construct reality has also been discussed in visual theory in light of the visual turn. See Horst Bredekamp, *Theorie des Bildakts* (Frankfurt: Suhrkamp, 2010); W. J. T. Mitchell, *What Do Pictures Want? The Lives and Loves of Images* (Chicago: University of Chicago Press, 2005).

33. Paglen, "Invisible Images," 2.

34. See e.g., Lorraine Daston and Peter Galison, "The Image of Objectivity," *Representations* 40 (1992): 81–128.

35. Paglen, "Invisible Images," 8.

36. Simone Browne, *Dark Matters: On the Surveillance of Blackness* (Durham, NC: Duke University Press, 2015).

37. "Interestingly, when their gender classifier was made 'ethnicity specific' for the category 'African,' they found that images of African females would be classified as

female about 82 percent of the time, while the same African classifier would find images of 'Mongoloid' females to be female 95.5 percent of the time, and 96 percent for 'Caucasoid' females." Browne, *Dark Matters*, 111.

38. Joy Buolamwini and Timnit Gebru, "Gender Shades: Intersectional Accuracy Disparities in Commercial Gender Classification," *Proceedings of Machine Learning Research* 81 (2018): 77–91.

39. For a discussion of Zach Blas's art, see Lee-Morrison, *Portraits*, 141–157.

40. Aaron L. Mishara, "Klaus Conrad (1905–1961): Delusional Mood, Psychosis, and Beginning Schizophrenia," *Schizophrenia Bulletin* 36, no. 1 (2010): 9–13.

41. Matteo Meschiari, "Roots of the Savage Mind: Apophenia and Imagination as Cognitive Process," *Quaderni di Semantica* 2, no. 1 (2009): 1–39.

42. Hito Steyerl, "A Sea of Data: Apophenia and Pattern (Mis-)recognition," *e-flux journal* 72 (2016): 1–14.

43. Steyerl, "A Sea of Data," 2.

44. Steyerl, "A Sea of Data," 9.

45. Steyerl, "A Sea of Data," 2.

46. On the poetization of clouds in German Romanticism and realism, see Kathrin Maurer, "Adalbert Stifter's Poetics of Clouds and Nineteenth-Century Meteorology," *Oxford German Studies* 45, no. 4 (2016): 421–433.

47. Walter Benjamin, "On Some Motifs in Baudelaire," in *Illuminations: Essays and Reflections*, ed. Hannah Arendt (New York: Schocken, 1968), 188.

48. Rosi Braidotti, *The Posthuman* (Cambridge: Polity, 2013), 57.

49. Matt McFarland, "Slaughterbots Film Shows Potential Horrors of Killer Drones," *CNNMoney*, November 14, 2017, https://money.cnn.com/2017/11/14/technology /autonomous-weapons-ban-ai/index.html.

50. *Slaughterbots*, directed by Stewart Sugg, 2017, 7:47 mins., https://www.youtube .com/watch?v=9CO6M2HsoIA.

51. I am indebted to Jutta Weber, who drew my attention to the movie *Slaughterbots* and whose research was an inspiration for my writing about the film. Jutta Weber, "Artificial Intelligence and the Socio-Technical Imaginary: On Skynet, Self-Healing Swarms and *Slaughterbots*," in *Drone Imaginaries: The Remote Power of Vision*, ed. Andreas Immanuel Graae and Kathrin Maurer (Manchester: Manchester University Press, 2021), 167–179.

52. *Slaughterbots*, 0:48 mins.

53. *Slaughterbots*, 2:15 mins.

54. Irma van der Ploeg, "Biometrics and the Body as Information," in *Surveillance as Social Sorting: Privacy, Risk, and Digital Discrimination*, ed. David Lyon (New York: Routledge, 2003), 57–73.

55. Zygmunt Bauman and David Lyon, *Liquid Surveillance: A Conversation* (Hoboken: John Wiley & Sons, 2013).

56. Lyon, *Surveillance as Social Sorting*.

57. David Lyon, "Surveillance and Social Sorting: Computer Codes and Social Sorting," in Lyon, *Surveillance as Social Sorting*, 13–20.

58. "It classifies and categorizes relentlessly, on the basis of various—clear or occluded—criteria." David Lyon, "Introduction," in Lyon, *Surveillance as Social Sorting*, 8.

59. Lee-Morrison, *Portraits*, 85–100; Gates, *Our Biometric Future*, 19.

60. *Slaughterbots*, 1:51 mins.

61. On Skynet, see Weber, "Artificial Intelligence," 167–179.

62. Paul Scharre, "Why You Should Not Fear Slaughterbots," *IEEE Spectrum*, December 22, 2017, https://spectrum.ieee.org/automaton/robotics/military-robots/why-you-shouldnt-fear-slaughterbots.

63. Stuart Russell et al., "Why You Should Fear Slaughterbots," *IEEE Spectrum*, January 23, 2018, https://spectrum.ieee.org/automaton/robotics/artificial-intelligence/why-you-should-fear-slaughterbots-a-response.

64. Adam Harvey, AH Projects, accessed October 2, 2021, https://ahprojects.com. On *CV Dazzle* specifically, see https://ahprojects.com/cvdazzle.

65. This algorithm is already somewhat old-fashioned, and *CV Dazzle* would not work against deep convolutional neural networks. See https://ahprojects.com/cvdazzle.

66. I owe this insight Svea Braeunert, "To See without Being Seen: Contemporary Art and Drone Warfare," in *To See without Being Seen: Contemporary Art and Drone Warfare*, ed. Svea Braeunert and Meredith Malone (Chicago: University of Chicago Press, 2016), 24.

67. Adam Harvey, AH Projects.

68. Lisa Parks, "Drones, Infrared Imagery, and Body Heat," *International Journal of Communication* 8 (2014): 2518–2521.

69. Quoted in Jennifer Rhee, "Adam Harvey's 'Anti-Drone' Wear in Three Sites of Opacity," *Camera Obscura: Feminism, Culture, and Media Studies* 31, no. 2 (2016): 180.

70. Rhee also problematizes Harvey's statements about the burqa. By emphasizing the burqa's empowering aspects, he is arguably caught in a culturally essentializing

and reductive perspective, and his aestheticization of the burqa suggests a reduction and an essentialization of Muslim culture more widely. Rhee, "Adam Harvey's 'Anti-Drone' Wear," 180.

71. See Mace Lee, "Stealth," *Urban Dictionary*, September 5, 2007, https://www.urbandictionary.com/define.php?term=stealth.

72. Shoshana Magnet, *When Biometrics Fail: Gender, Race, and the Technology of Identity* (Durham, NC: Duke University Press, 2011), 48.

73. Magnet, *When Biometrics Fail*, 50.

74. Braeunert, "To See without Being Seen," 23.

75. See also Göksu Kunak, "Interview: Hito Steyerl: Zero Probability and the Age of Mass Art Production," *Berlin Art Link*, November 19, 2013, http://www.berlinartlink.com/2013/11/19/interview-hito-steyerl-zero-probability-and-the-age-of-mass-art-production/.

76. Hito Steyerl and Laura Poitras, "Techniques of the Observer: Hito Steyerl and Laura Poitras in Conversation," *Artforum International* 53, no. 9 (2015): 308. I owe the discovery of this quote to Braeunert, "To See without Being Seen," 23.

77. On the idea that surveillance can also be conducted from below, see Steve Mann, Jason Nolan, and Barry Wellman, "Sousveillance: Inventing and Using Wearable Computing Devices for Data Collection in Surveillance Environments," *Surveillance & Society* 1, no. 3 (2003): 331–355. On an idea that inspired artists and activists, see Jan Fernback, "Sousveillance: Communities of Resistance to the Surveillance Environment," *Telematics and Informatics* 30, no. 1 (2013): 11–21.

Chapter 4

1. *5000 Feet Is the Best*, directed by Omer Fast, 2011, 25:09 mins.

2. On flattening and the military drone view, see Michael Andreas, "Flächen/Rastern: Zur Bildlichkeit der Drohne," *Behemoth: A Journal on Civilisation* 8, no. 2 (2015): 108–127.

3. For research on the notion of the planetary within the humanities, see Christian Moraru, *Reading for the Planet: Toward a Geomethodology* (Ann Arbor: University of Michigan Press, 2015); Masao Miyoshi, "Turn to the Planet: Literature, Diversity, and Totality," *Comparative Literature* 53, no. 4 (2001): 283–297; Ursula Heise, *Sense of Place and Sense of Planet: The Environmental Imagination of the Global* (New York: Oxford University Press, 2010); Amy J. Elias and Christian Moraru, eds., *The Planetary Turn: Relationality and Geoaesthetics in the Twenty-First Century* (Evanston: Northwestern University Press, 2015).

4. Gayatri Chakravorty Spivak, "Planetarity," in *Death of a Discipline* (New York: Columbia University Press, 2003), 71–102.

5. Spivak, "Planetarity," 72.

6. The discourse of planetarity is closely entwined with that of the Anthroprocene, on which I expand in chapter 5.

7. Heather McLean, "In Praise of Chaotic Research Pathways: A Feminist Response to Planetary Urbanization," *Environment and Planning: Society and Space* 36, no. 3 (2018): 547–555.

8. I have written elsewhere about the relationship between drones and balloons: Kathrin Maurer, "Flattened Vision: Nineteenth-Century Hot Air Balloons as Early Drones," in *Drone Imaginaries: The Remote Power of Vision*, ed. Andreas Immanuel Graae and Kathrin Maurer (Manchester: Manchester University Press, 2021), 19–38. In this chapter, I trace the similarity between drones and hot-air balloons, but I focus mainly on literary representations of aerial balloons in German Romanticism, and therefore the text differs from the one in this chapter. See also Kathrin Maurer, "Ballooning as a Technology of Seeing in Jean Paul's *Des Luftschiffers Giannozzo Seebuch* (1801)," in *Before Photography*, ed. Kirsten Belgum, Vance Byrd, and John D. Benjamin (Berlin: Walter de Gruyter, 2021), 17–38.

9. On Nadar's work as a photographer and balloonist, see Adam Begely, *The Great Nadar: The Man behind the Camera* (New York: Penguin Random House, 2017).

10. On aerial balloons in popular culture, see Alfred Eckert, "Zur Geschichte der Ballonfahrt," in *Leichter als Luft: Zur Geschichte der Ballonfahrt*, ed. Bernard Korzus and Gisela Noehles (Münster: Westfälisches Landesmuseum für Kunst und Kulturgeschichte, 1978), 13–128.

11. For an overview of literary works that engage with the aerial balloon, see Eckhard Schinkel, "Der Ballon in der Literatur," in Korzus and Noehles, *Leichter als Luft*, 200–236.

12. Caren Kaplan, "Drone-o-Rama: Troubling the Temporal and Spatial Logics of Distance Warfare," in *Life in the Age of Drone Warfare*, ed. Caren Kaplan and Lisa Parks (Durham, NC: Duke University Press, 2017), 161–177.

13. Quoted in John Christopher, *Balloons at War: Gasbags, Flying Bombs, and Cold War Secrets* (Stroud: Tempus, 2004), 49.

14. Thomas Baldwin, *Airopaidia* (London: J. Fletcher, 1786). I am indebted to Caren Kaplan's work on aerial balloons and drones, which brought my attention to Baldwin. But in contrast to Kaplan, I emphasize the nonmilitary and planetary dimension of early ballooning. See Caren Kaplan, "Balloon Geography: The Emotion of Motion in Aerostatic Wartime," in *Aerial Aftermaths: Wartime from Above* (Durham, NC: Duke University Press, 2017), 68–103.

15. Baldwin, *Airopaidia*, 38.

16. On the criteria of operative visuality, see Sybille Krämer, "Operative Bildlichkeit: Von der Grammatologie zu einer Diagrammatologie? Reflexionen über erkennendes Sehen," in *Logik des Bildlichen: Zur Kritik der ikonischen Vernunft*, ed. Martina Hessler and Dieter Mersch (Bielefeld: Transcript, 2009), 94–122.

17. Gottfried Boehm, "Zwischen Auge und Hand: Bilder als Instrumente der Erkenntnis," in *Mit dem Auge Denken: Strategien der Sichtbarmachung in wissenschaftlichen und virtuellen Welten*, ed. Bettina Heintz and Jorg Huber (Vienna: Springer, 2001), 43–54; Stefan Majetschak, "Sichtvermerke: Über den Unterschied zwischen Kunst und Gebrauchsbildern," in *Bild-Zeichen: Perspektiven einer Wissenschaft vom Bild*, ed. Stefan Majetschak (Munich: Fink, 2005), 97–121.

18. Krämer, "Operative Bildlichkeit," 98.

19. Krämer, "Operative Bildlichkeit," 98.

20. Andreas, "Flächen/Rastern."

21. "In this sense, I take some distance, so to speak, from the work of Lisa Parks, whose *Cultures in Orbit: Satellites and the Televisual* (Durham: Duke University Press, 2005) is an excellent study of the culture of satellite images in politics, the news, and our imagination. Her interpretations, however, stop at the ways in which these images are presented to us, which is to say that she reads only what has already been interpreted." Laura Kurgan, *Close Up at a Distance: Mapping, Technology, and Politics* (New York: Zone, 2013), 217.

22. Svetlana Alpers, *The Art of Describing: Dutch Art in the Seventeenth Century* (Chicago: University of Chicago Press, 1984), 141.

23. On the aesthetics of the ornament and the drone, see Christoph Asendorf's discussion of Kazimir Malevich, the aerial view of flattening, and avant-garde aesthetics: Christoph Asendorf, "Bewegliche Fluchtpunkte: Der Blick von oben und die moderne Raumanschauung," in *Iconic Worlds: Neue Bilderwelten und Wissensräume*, ed. Christa Maar and Hubert Burda (Cologne: DuMont, 2006), 11–16. See also Jan Mieszkowski, "The Drone of Data," in Graae and Maurer, *Drone Imaginaries*, 55–73. For more on Kracauer, mass ornament, and drones, see chapter 7.

24. Baldwin, *Airopaidia*, 2, 23, 38.

25. Baldwin, *Airopaidia*, 16.

26. Baldwin, *Airopaidia*, 3.

27. Edmund Burke, *Philosophical Enquiry into the Origin of Our Ideas of the Sublime and Beautiful* (Oxford: Oxford University Press, 2015), first published 1757.

28. Baldwin, *Airopaidia*, 109.

29. Baldwin, *Airopaidia*, 53.

30. Baldwin, *Airopaidia*, 38.

31. Baldwin, *Airopaidia*, 3.

32. Joanna Zylinska, "Photography after Extinction," in *After Extinction*, ed. Richard Grusin (Minneapolis: University of Minnesota Press, 2018), 51–70.

33. Baldwin, *Airopaidia*, 40.

34. Bruno Latour, *Facing Gaia: Eight Lectures on the New Climatic Regime* (Cambridge, UK: Polity, 2017), 94.

35. I am aware, however, that the notion of Gaia can also be connected to a model of holism and totality—a model Latour detests, and which I also wish to resist: Bruno Latour, "Why Gaia Is Not a God of Totality," *Theory, Culture & Society* 34, nos. 2–3 (2017): 61–81. For a critical discussion of Gaia as a premodern trope, see e.g., Daniele Sands, "Gaia Politics, Critique, and the Planetary Imaginary," *Substance* 49, no. 3 (2020): 104–121.

36. Trevor Paglen, *I Could Tell You but Then You Would Have to Be Destroyed by Me: Emblems from the Pentagon's Dark World* (Brooklyn: Melville House, 2006).

37. Andreas, "Flächen/Rastern."

38. Bernd Siegert, "(Not) in Place: The Grid, or, the Cultural Techniques of Ruling Spaces," in *Cultural Techniques: Grids, Filters, Doors, and Other Articulations of the Real* (New York: Fordham University Press, 2015), 97–120. Another important discussion is provided by Rosalind Krauss, who debates the grid from an art historical perspective and as a signature of the avant-garde: Rosalind Krauss, "Grids," *October* 9 (1979): 51–64.

39. Siegert, "(Not) in Place," 98.

40. On the connection between war and algorithms, see Louise Amoore, "Algorithmic War: Everyday Geographies of the War on Terror," *Antipode* 41, no. 1 (2009): 49–69. Amoore shows how the increase in security surveillance via algorithms domesticizes war, and what that means for our societies and communities. See also Max Liljefors, Gregor Noll, and Daniel Steuer, eds., *War and Algorithm* (London: Rowman and Littlefield, 2020).

41. See also my discussion of Paglen's *Drone Vision* in Kathrin Maurer, "Drones as Big Data Archives: Mimesis and Counter-Archiving in Contemporary Art on Military Drones," in *(W)archives*, ed. Daniela Agostinho et al. (Berlin: Sternberg Press, 2020), 119–141.

42. Terry Smith, "Comparing Contemporary Arts; or, Figuring Planetarity," in Elias and Moraru, *The Planetary Turn*, 184.

43. Smith, "Comparing Contemporary Arts," 184.

44. Quoted in Derek Gregory, "Seeing Machines," *Geographical Imaginations: Wars, Spaces and Bodies* (blog), April 15, 2014, https://geographicalimaginations.com/2014 /04/15/seeing-machines/.

45. Moraru, *Reading for the Planet*, 74.

46. Jessica L. Horton, "Drones and Snakes," *Art in America* 105, no. 9 (2017): 104–109.

47. "About," Tomas van Houtryve, accessed June 2, 2021, https://tomasvh.com/about/.

48. Tomas van Houtryve and Svea Braeunert, "Empathy and the Image under Surveillance Capitalism: Interview with Photographer Tomas van Houtryve," in Graae and Maurer, *Drone Imaginaries*, 74–88.

49. Hester Blum, "Terraqueous Planet: The Case for Oceanic Studies," in Elias and Moraru, *The Planetary Turn*, 25–36.

50. Houtryve and Braeunert, "Empathy and the Image," 82.

51. John Pizer, "Planetary Poetics: World Literature, Goethe, Novalis, and Yoko Tawada's Translational Writing," in Elias and Moraru, *The Planetary Turn*, 3–24. See also my work on water metaphors in transnational literature: Kathrin Maurer, "Translating Catastrophes: Yoko Tawada's Poetic Responses to the 2011 Tōhoku Earthquake, the Tsunami, and Fukushima," *New German Critique* 43, no. 1 (2016): 171–194.

52. Spivak, "Planetarity," 72.

53. Caren Kaplan and Andrea Miller, "Drones as Atmospheric Policing: From US Border Enforcement to the LAPD," *Public Culture* 31, no. 3 (2019): 419–445; Anna Feigenbaum and Anja Kanngieser, "For a Politics of Atmospheric Governance," *Dialogues in Human Geography* 5, no. 1 (2015): 80–84.

54. I owe my knowledge of *The Repellent Fence* to Caren Kaplan, who introduced me to Postcommodity's work when she gave a lecture at the University of Southern Denmark in 2017, and to her chapter "Eyes in the Skies: *Repellent Fence* and Trans-Indigenous Time-Space at the US-Mexico Border," in Graae and Maurer, *Drone Imaginaries*, 203–224.

55. Kate Morris, *Shifting Grounds: Landscape in Contemporary Native American Art* (Seattle: Washington University Press, 2019), 102.

56. *Through the Repellent Fence*, directed by Sam Wainwright Douglas, 2017.

57. Postcommodity, accessed October 3, 2021, http://postcommodity.com/.

58. Kaplan, "Eyes in the Skies," 203–224.

59. Mark Trecka, "The Implication of a Fence: Part Three: The Sovereignty of Context," *Beacon Broadside* (blog), July 6, 2016, https://www.beaconbroadside.com/broadside /2016/07/the-implication-of-a-fence-part-three-the-sovereignty-of-context.html.

60. On discourses of stewardship in planetary studies, see Moraru, *Reading for the Planet*, 174–181; Timothy Luke, "On the Politics of the Anthropocene," *Telos* 172 (2015): 139–162; Kennedy Graham, ed., *The Planetary Interest: A New Concept for the Global Age* (New Brunswick: Rutgers University Press, 1999).

61. Rosi Braidotti, *Posthuman Knowledge* (Cambridge, UK: Polity, 2019), 52.

62. Kaplan, "Eyes in the Skies," 203–224.

63. See also Austin Choi-Fitzpatrick, *The Good Drone: How Social Movements Democratize Surveillance* (Cambridge, MA: MIT Press, 2020).

64. See Kaplan's reference to Andrea Miller's work on the atmosphere in respect to *The Repellent Fence*: Kaplan, "Eyes in the Skies," 216.

65. Ignacio Acosta, *Copper Geographies* (Barcelona: Editorial RM, 2018).

66. Neil Kent, *The Sami Peoples of the North: A Social and Cultural History* (New York: Hurst & Co., 2018).

67. In fact, there is a wide variety of drone images in *Drones and Drums*. The drone films inside the forest and takes panoramic and first-person view images, as well as detailed images of trees, birds, and plants. The film is not exclusively made by drone, but also uses other camera types.

68. For research on eco-media, see Gay Hawkins, *The Ethics of Waste: How We Relate to Rubbish* (Lanham: Rowman and Littlefield, 2005), 113; Sean Cubitt, *EcoMedia* (Amsterdam: Rodopi, 2005); Richard Maxwell and Toby Miller, *Greening the Media* (Oxford: Oxford University Press, 2012); Jennifer Gabrys, *Digital Rubbish: A Natural History of Electronics* (Ann Arbor: University of Michigan Press, 2011); Jussi Parikka, *A Geology of Media* (Minneapolis: University of Minnesota Press, 2015); Jussi Parikka, *Medianatures* (London: Open Humanities Press, 2013).

69. See Choi-Fitzpatrick, *The Good Drone*.

70. Sarah Tuck, "Drone Alliances," in *Fragmentation of the Photographic Image in the Digital Age*, ed. Daniel Rubenstein (New York: Routledge, 2020), 73–79; Horton, "Drones and Snakes."

71. Tuck, "Drone Alliances," 77.

Chapter 5

1. Jennifer Gabrys, *Program Earth: Environmental Sensing Technology and the Making of a Computational Planet* (Minneapolis: University of Minnesota Press, 2016), 7.

2. Ole B. Jensen, "Thinking with the Drone: Visual Lessons in Aerial and Volumetric Thinking," *Visual Studies* 35, no. 5 (2020): 1.

3. Jensen, "Thinking with the Drone," 418.

4. Jensen, "Thinking with the Drone"; Franciso Klauser, "Splintering Spheres of Security: Peter Sloterdijk and the Contemporary Fortress City," *Environment and Planning: Society and Space* 28, no. 2 (2010): 326–340; Francisco Klauser, *Surveillance and Space* (London: Sage, 2016); Stuart Elden, "Secure the Volume: Vertical Geopolitics and the Depth of Power," *Political Geography* 1, no. 34 (2013): 35–51. Although Stephen Graham works on verticality, his critique of flatness connects verticality to the volumetric: "In practice, this means that human societies are increasingly dense and stacked societies, in which uses of space are built upwards and downwards with ever-greater intensity within geographical volumes." Stephen Graham, *Vertical: The City from Satellites to Bunkers* (London: Verso, 2016), 4.

5. For an early discussion of the Anthropocene, see Elizabeth Kolbert, "Enter the Anthropocene Age of Man," *National Geographic* 219, no. 3 (2011): 60–85.

6. For theoretical discussions of the Anthropocene, see Christophe Bonneuil, Jean-Baptiste Fressoz, and David Fernbach, eds., *The Shock of the Anthropocene: The Earth, History and Us* (London: Verso, 2016). For a discussion of the notion of the Anthropocene in the humanities, see e.g., David Farrier, *Anthropocene Poetics: Deep Time, Sacrifice Zones, and Extinction* (Minneapolis: University of Minnesota Press, 2019).

7. Timothy Morton, *Ecology without Nature: Rethinking Environmental Aesthetics* (Cambridge, MA: Harvard University Press, 2007).

8. See e.g., Bruno Basso, James Dobrowski, and Channing McKay, "From the Dust Bowl to Drones to Big Data: The Next Revolution in Agriculture," *Georgetown Journal of International Affairs* 18, no. 3 (2017): 158–165.

9. Although cultural surveillance studies is more geared toward the urban, there has been some research that investigates smart farming from the perspective of power, surveillance, and control. I owe my knowledge about agricultural drones in a surveillance studies context to the following: Francisco Klauser, "Surveillance Farm: Towards a Research Agenda on Big Data Agriculture," *Surveillance & Society* 16, no. 3 (2018): 370–378; Brad Bolman, "A Revolution in Agricultural Affairs: Dronoculture, Precision, Capital," in Kristin Bergtora Sandvik and Maria Gabrielsen Jumbert, eds., *The Good Drone* (New York: Routledge, 2016), 129–151; Brad Bolman, "Provocation: A Prairie Drone Companion," *Culture Machine* 16 (2015): 1–6.

10. Bolman, "A Revolution," 129.

11. "Drone Software & Technology Providers," Postscapes, accessed May 2021, https://www.postscapes.com/agriculture-drone-companies/#software.

12. Gabrys, *Program Earth*, 3.

13. Gabrys, *Program Earth*, 4.

14. Chein-I Chang, *Hyperspectral Imaging: Techniques for Spectral Detection and Classification* (New York: Kluwer Academic/Plenum, 2003).

15. Lisa Parks, "Drones, Infrared Imagery, and Body Heat," *International Journal of Communication* 8 (2014): 2518–2521.

16. Charles Veys et al., "An Ultra-Low-Cost Active Multispectral Crop Diagnostics Device," in *2017 IEEE Sensors (Proceedings)*, ed. Krikor Oznayan (New York: Institute of Electrical and Electronics Engineers, 2017), 1005–1008.

17. Hilde Schoofs et al., "Fire Blight Monitoring in Pear Orchards by Unmanned Airborne Vehicles (UAV) Systems Carrying Spectral Sensors," *Agronomy* 10, no. 5 (2020): 1–12.

18. Flor Alvarez-Taboada, Claudio Paredes, and Julia Julián-Pelaz, "Mapping of the Invasive Species *Hakea sericea* Using Unmanned Aerial Vehicle (UAV) and WorldView-2 Imagery and an Object-Oriented Approach," *Remote Sensing* 9, no. 9 (2017): 1–17.

19. Alvarez-Taboada, Paredes, and Julián-Pelaz, "Mapping," 2.

20. Alvarez-Taboada, Paredes, and Julián-Pelaz, "Mapping," 3.

21. Grégoire Chamayou, *A Theory of the Drone* (New York: New Press, 2015).

22. Chamayou, *Theory of the Drone*, 146.

23. Kathrin Maurer, "Visual Power: The Scopic Regime of Military Drones," *War, Media, Conflict* 10, no. 2 (2017): 141–151.

24. Bolman, "A Revolution," 143.

25. Eva Horn, "Challenges for an Aesthetics of the Anthropocene," in *The Anthropocenic Turn: The Interplay between Disciplinary and Interdisciplinary Responses to a New Age*, ed. Gabriele Dürbeck and Philip Hüpkes (New York: Routledge, 2020), 97–111. See also Nicholas Mirzoeff's critique of the aestheticization of the Anthropocene: Nicholas Mirzoeff, "Visualizing the Anthropocene," *Public Culture* 26, no. 2 (2014): 213–232.

26. One of his recent works is the multimedia project *The Anthropocene*, which consists of a photo series, a book, and the film *Anthropocene: The Human Epoch* (2018). Edward Burtynsky. Jennifer Baichwal, and Nicolas de Pencier, *Anthropocene* (Göttingen: Steidl, 2018). In chapter 7, I also discuss his New York City drone photography in relation to the COVID-19 pandemic.

27. I am thankful to Lila Lee-Morrison for introducing me to George Steinmetz's work.

28. George Steinmetz, "Drones Are Changing How We See the World," *Time*, May 31, 2018, https://time.com/longform/drones-career.

29. George Steinmetz and Andrew Revkin, *The Human Planet: Earth at the Dawn of the Anthropocene* (New York: Abrams, 2020).

30. Steinmetz and Revkin, *The Human Planet*, 250.

31. This is particularly interesting in the context of air warfare: Peter Sloterdijk, *Terror from the Air*, trans. Amy Patton and Steve Corcoran, (Los Angeles: Semiotext(e), 2009); Caren Kaplan, "Atmospheric Politics: Protest Drones and the Ambiguity of Airspace," *Digital War* 1 (2020): 50–57. In this chapter (compared with chapter 4), the idea of atmospheric space is mostly connected to the discourse of the Anthropocene. I remain aware that the atmosphere has both military implications and vital, life-sustaining dimensions.

32. Timothy Morton, *Hyperobjects: Philosophy and Ecology after the End of the World* (Minneapolis: University of Minnesota Press, 2013).

33. Timothy Morton, *The Ecological Thought* (Cambridge, MA: Harvard University Press, 2010), 14.

34. On drones and glacier observation, see Kristaps Lamsters et al., "Application of Unmanned Aerial Vehicles for Glacier Research in the Arctic and Antarctic," *Environment, Technologies, Resources: Proceedings of the International Scientific and Practical Conference* 1 (2019): 131–135. On wildlife conservation, see Serge Wich, Lorna Scott, and Lian Pin Koh, "Wings for Wildlife: The Use of Conservation Drones, Challenges and Opportunities," in Sandvik and Jumbert, *The Good Drone*, 163–177.

35. J. D. Schnepf, "Flood from Above: Disaster Mediation and Drone Humanitarianism," *Media + Environment* 2, no. 1 (2020): 13466.

36. The work is in progress, and I had a glimpse of it at the symposium Drone Cultures: Interdisciplinary Perspectives, University of New South Wales, Sydney, December 8, 2020.

37. See also "Work in Progress," Senses of Perception, accessed November 15, 2022, https://sensesofperception.info/work-in-progress/.

38. Anna Munster and Michele Barker, "Ecologies of the Drone" (conference presentation, Drone Cultures: Interdisciplinary Perspectives, University of New South Wales, Sydney, December 8, 2020). See also Erin Manning, *Politics of Touch: Sense, Movement, Sovereignty* (Minneapolis: University of Minnesota Press, 2007).

39. Maria Puig de la Bellacasa, *Matters of Care: Speculative Ethics in More Than Human Worlds* (Minneapolis: Minnesota University Press, 2017).

40. Paul Henry Cureton, *Drone Futures: UAS in Landscape and Urban Design* (London: Routledge, 2021).

41. On the connection between technology, holism, and ecological imperialism, see Peder Anker, *Imperial Ecology: Environmental Order in the British Empire, 1895–1945* (Cambridge, MA: Harvard University Press, 2001).

42. Laura Kurgan, *Close Up at a Distance: Mapping, Technology, and Politics* (New York: Zone, 2013).

43. Nadia Amoroso, *The Exposed City: Mapping the Urban Invisibles* (New York: Routledge, 2010).

44. Karl Kullmann, "The Drone's Eye: Applications and Implications for Landscape Architecture," *Landscape Research* 43, no. 7 (2018): 906–921.

45. Brett Milligan, "Making Terrains: Surveying, Drones and Media Ecology," *Journal of Landscape Architecture* 14, no. 2 (2019): 22.

46. Rikke Munck Petersen's work highlights the sensorial experientiality of the drone in the context of urban and landscape planning. She shows that the drone enables an embodied perspective onto the earth, connecting the pilot/planner with the ground by sensually (visually, kinetically, and aurally) immersing them in the environment. Rikke Munck Petersen, "The Dispatched Drone and Affective Distance in Fieldwork," *The Senses and Society* 15, no. 3 (2020): 311–328.

47. Milligan, "Making Terrains," 20.

48. Milligan, "Making Terrains," 20.

49. Kullmann, "The Drone's Eye," 12.

50. Christophe Girot and James Melsom, "The Return of the Aviators," *Topos* 86 (2014): 102–107; Christophe Girot et al., "Scales of Topology in Landscape Architecture," *Scales of the Earth* 4 (2011): 156–163; Christophe Girot, "Cloudism: Towards a New Culture of Making Landscapes," in *Routledge Research Companion to Landscape Architecture*, ed. Ellen Braae and Henriette Steiner (London: Routledge, 2018), 113–123.

51. For more information on robotic landscapes and on Girot's work, see Prof. Christophe Girot, accessed October 2021, https://girot.arch.ethz.ch/tag/robotic-landscapes.

52. Denis Cosgrove and William L. Fox, *Photography and Flight* (London: Reaktion, 2010), 24, 26, 31, 56–57.

53. See Keith Armstrong, accessed October 3, 2021, http://embodiedmedia.com.

54. Kullmann critiques Girot's and James Melsom's work as apolitical and technical: "On account of its novelty, extant literature that directly addresses the topic of drone-based imaging and mapping in landscape architecture is primarily limited to professional and non-critical contexts (see Girot and Melsom 2014)." Kullmann, "The Drone's Eye," 4.

55. Keith Armstrong, "Embodying a Future for the Future: Creative Robotics and Ecosophical Praxis," *Fibreculture Journal* 28 (2016): 2.

56. Keith Armstrong and Tania Leimbach, "Art-Eco-Science: Field Collaborations," *Antennae: The Journal of Nature in Visual Culture* 48 (2019): 110.

57. Armstrong and Leimbach, "Art-Eco-Science," 110

58. Armstrong is influenced by Morton's *The Ecological Thought* and Arne Næss and Bob Jickling, "Deep Ecology and Education: A Conversation with Arne Naess," *Canadian Journal of Environmental Education* 5, no. 1 (2000): 48–62.

59. See Keith Armstrong, accessed October 3, 2021, http://embodiedmedia.com/.

60. Armstrong and Leimbach, "Art-Eco-Science," 125.

61. The exhibit can be seen at "Change Agent," Keith Armstrong, accessed October 18, 2021, http://embodiedmedia.com/homeartworks/change-agent.

62. Keith Armstrong, *Future-Future?* (Mount Gambier: Riddoch Art Gallery, 2017), https://www.vrystaatkunstefees.co.za/wp-content/uploads/2020/05/Future-Future -2017.pdf.

63. One finds a very different take on lidar software and remote vision in Liam Young's work on human exclusion zones, which issues warnings about machinic urban planning and total surveillance. Liam Young, *Machine Landscapes: Architectures of the Post Anthropocene* (Hoboken: John Wiley & Sons, 2019); Benjamin Bratton, "Further Trace Effects of the Post-Anthropocene," *Architectural Design* 89, no. 1 (2019): 14–21.

64. Adam Fish, "Drones at the Edge of Naturecultures," *Media Fields* 14 (2019): 1–5.

65. Cureton, *Drone Futures*, 198.

66. Bernard Stiegler, *The Neganthropocene* (London: Open Humanities Press, 2018).

Chapter 6

1. Drones raise a debate about autonomy, as Lucy Suchman and Jutta Weber's research exemplifies. Writing about military drones, Suchman and Weber differentiate between two traditions of autonomy. On the one hand, autonomy is an Enlightenment concept that had its beginnings in Immanuel Kant's moral philosophy: autonomy is an individual's capacity for self-determination or self-governance. On the other hand, this liberal model of autonomy is challenged by twentieth-century systems theory and cybernetics, which understand humans and nonhumans alike as autopoietic, self-regulating systems. If applied to drones, this technoscientific vision of autonomy takes the human somewhat out of the loop, and Suchman and Weber argue that in doing so it deprives us of ethical responsibility. In respect to drones in particular, it takes away our ability to interject, modify, stop, and steer machinic operations. Lucy Suchman and Jutta Weber, "Human-Machine Autonomies," in *Autonomous Weapon Systems: Law, Ethics, Policy*, ed. Bhuta Nehal et al. (New York: Cambridge University Press, 2016), 75–102. See also James Johnson, "Artificial Intelligence, Drone Swarming and Escalation Risks in Future Warfare," *RUSI Journal* 165, no. 2 (2020): 26–36.

2. For a discussion of drone swarms and swarming as a military strategy, see Paul Scharre, *Army of None: Autonomous Weapons and the Future of War* (New York: Norton, 2018); Paul Scharre, "How Swarming Will Change Warfare," *Bulletin of the Atomic Scientists* 74, no. 6 (2018): 385–389; John Arquilla and David Ronfeldt, *Swarming and the Future of Conflict* (Santa Monica: Rand, 2000); Francis Grimal and Jae Sundaram, "Combat Drones: Hives, Swarms, and Autonomous Action?" *Journal of Conflict and Security Law* 23, no. 1 (2018): 105–135; Goetz Herrmann and Jutta

Weber, "Game of Swarms: Swarm Technologies, Control, and Autonomy in Complex Weapons Systems," in *Game of Drones: Of Unmanned Aerial Vehicles*, ed. Claudia Emmert et al. (Berlin: Neofelis, 2020), 174–196. Indeed, as 2021's drone missions in Nagorno-Karabakh, Azerbaijan, indicate, drone swarms are already being implemented in military practice: Jason Crabtree, "Gaza and Nagorno-Karabakh Were Glimpses of the Future of Conflict," *FP*, June 21, 2021, https://foreignpolicy.com/2021/06/21/gaza-nagorno-karabakh-future-conflict-drones/.

3. On the development and implementation of civilian drone swarms, see Anam Tahir et al., "Swarms of Unmanned Aerial Vehicles: A Survey," *Journal of Industrial Information Integration* 16 (2019): 100106; Balsam Alkouz, Athman Bouguettaya, and Sajib Mistry, "Swarm-Based Drone-as-a-Service (SDAAS) for Delivery," in *2020 IEEE International Conference on Web Services Proceedings*, ed. Lisa O'Connor (Los Alamitos: Institute of Electrical and Electronic Engineers, 2020), 441–448.

4. South China Morning Post, *Shanghai's Drone Show Welcoming 2020 Reportedly Never Happened on New Year's Eve*, YouTube, 2:25, uploaded January 30, 2020, https://www.youtube.com/watch?v=F_DkUXiLczE.

5. Cirque du Soleil, *Sparked: Behind the Technology*, YouTube, 2:39, uploaded September 22, 2014, https://www.youtube.com/watch?v=7YqUocVcyrE.

6. *Black Mirror*, "Hated in the Nation," episode directed by James Hawes, released 2016 on Netflix, 89 mins.

7. Ernst Jünger, *The Glass Bees* (New York: Noonday Press, 1960), first published 1957 in German.

8. Eva Horn and Lucas Marco Gisi, eds., *Schwärme: Kollektive ohne Zentrum: Eine Wissensgeschichte zwischen Leben und Information* (Bielefeld: Transcript, 2009).

9. Horn, "Einleitung," in Horn and Gisi, *Schwärme*, 7–16.

10. Alexander R. Galloway and Eugene Thacker, *The Exploit: A Theory of Networks* (Minneapolis: University of Minnesota Press, 2013).

11. Galloway and Thacker, *The Exploit*, 5.

12. Jussi Parikka, "Politics of Swarms: Translations between Entomology and Biopolitics," *Parallax* 14, no. 3 (2008): 115. The relationship between ambivalence and the uncanny is famously explored by Sigmund Freud, as well as in Tzvetan Todorov's theory of the fantastic as a state of epistemological ambivalence in Romantic literature.

13. Andreas Immanuel Graae and Kathrin Maurer, "Introduction," in *Drone Imaginaries: The Remote Power of Vision*, ed. Andreas Immanuel Graae and Kathrin Maurer (Manchester: Manchester University Press, 2021), 1.

14. Jeffrey Lockwood, *The Infested Mind: Why Humans Fear, Loathe, and Love Insects* (Oxford: Oxford University Press, 2013).

15. Horn, "Einleitung," 20.

16. My discussion is indebted to recent research that explicitly investigates the inhuman and dystopian aspects of the drone swarm as a surveillance technology in "Hated in the Nation": Andreas Graae, "Swarming Sensations: Robo-bees and the Politics of the Swarm in *Black Mirror*," *The Senses and Society* 15, no. 3 (2020): 329–343; Garbor Sarlos, "Hated in the Nation: A Phantasma of the Post-Climate World," in *Reading Black Mirror: Insights into Technology and the Post-Media Condition*, ed. German A. Duarte and Justin Micheal Battin (Bielefeld: Transcript, 2021), 309–324.

17. Geoffrey R. Williams and David R. Tarpy, "Colony Collapse Disorder in Context," *Bioessays* 32, no. 10 (2010): 845; Jean-Marc Bonmatin et al. "Entomology: The Bee-All and End-All," *Nature* 521 (2015): 57–59.

18. Peter Soroye, Tim Newbold, and Jeremy Kerr, "Climate Change Contributes to Widespread Declines among Bumble Bees across Continents," *Science* 367, no. 6478 (2020): 685–688.

19. "Hated in the Nation," 32:36 mins.

20. "Hated in the Nation," 29:41–30:02 mins.

21. "Hated in the Nation," 32:01–32:18 mins.

22. Galloway and Thacker, *The Exploit*, 66. Galloway and Thacker place this in a military context: the swarm as faceless enemy. They refer to the 2003 film *The Matrix Revolutions*, in which swarms of insectile sentinels are eventually defeated by the humans' activity, which amasses the individual sentinels into one large "anthropomorphic face—a literal facialization of enmity" (69).

23. On swarm intelligence and social media, see Shuangling Luo et al., "Toward Collective Intelligence of Online Communities: A Primitive Conceptual Model," *Journal of Systems Science and Systems Engineering* 18, no. 2 (2009): 203–221.

24. Lauren Wilcox connects representations of swarms to drone warfare, but she also shows that insect swarming has been used as a community model in feminist sci-fi novels, and in feminist studies that adopt new materialist and nonhumanist perspectives. Besides dystopic visions of insect states, for example, she analyzes more positive representations of "becoming-insect" as feminist utopias in Charlotte Perkins Gilman's novel *Herland* (1915), and in *Houston, Houston: Do You Read?* (1976), a novella by James Tiptree, Jr. (a pseudonym of Alice Bradley Sheldon). Lauren Wilcox, "Drones, Swarms and Becoming-Insect: Feminist Utopias and Posthuman Politics," *Feminist Review* 116, no. 1 (2017): 25–45; see also Lauren Wilcox, "The Gender Politics of the Drone," in Graae and Maurer, *Drone Imaginaries*, 110–129. Rosi Braidotti's work on insects and their potential for rethinking sociocentrism through the lens of new materialism and feminism is also interesting in this context. For Braidotti, David Cronenberg's film *The Fly* (1986), in which the protagonist transforms into

an insect, represents a way of scrambling the master code of phallocentrism. Rosi Braidotti, *Metamorphoses: Towards a Materialist Theory of Becoming* (Hoboken: John Wiley & Sons, 2013), chapter 3.

25. Jussi Parikka, *Insect Media: An Archaeology of Animals and Technology* (Minneapolis: University of Minnesota Press, 2010), 39.

26. Parikka, "Politics of Swarms," 113.

27. I owe this reference to Parikka, *Insect Media*, 226.

28. Parikka, "Politics of Swarms," 3.

29. John Johnston, *The Allure of Machinic Life: Cybernetics, Artificial Life, and the New AI* (Cambridge, MA: MIT Press, 2008).

30. Norbert Wiener, *Cybernetics or Control and Communication in the Animal and the Machine* (Cambridge, MA: MIT Press, 2019), first published 1948.

31. Parikka, *Insect Media*, 123.

32. Parikka, *Insect Media*, 12; Graae, "Swarming Sensations," 329–343.

33. Parikka, *Insect Media*, 43.

34. Roger Berkowitz, "Drones and the Question of the Human," *Ethics & International Affairs* 28, no. 2 (2014): 159; Andreas Immanuel Graae, "Swarm of Steel: Insects, Drones and Swarming in Ernst Jünger's *The Glass Bees*," in Graae and Maurer, *Drone Imaginaries*, 149–166; Antoine Bousquet, *The Eye of War: Military Perception from the Telescope to the Drone* (Minneapolis: University of Minnesota Press, 2018).

35. For a biography of Ernst Jünger, see Helmuth Kiesel, *Ernst Jünger: Die Biographie* (Munich: Siedler, 2009). On Jünger's political positioning, see e.g., Elliot Neaman, *A Dubious Past: Ernst Jünger and the Politics of Literature after Nazism* (Berkeley: University of California Press, 1999). Andreas Huyssen offers a critical reading of Jünger's fascination with fascist ideology and aesthetics: Andreas Huyssen, "Fortifying the Heart—Totally: Ernst Jünger's Armored Texts," *New German Critique* 59, no. 1 (1993): 3–23.

36. For more on Jünger's inspiration by the Macy Conferences and Wiener's theory of cybernetics, see Niels Werber, "Ants and Aliens: An Episode in the History of Entomological and Sociological Construction of Knowledge," *Berichte zur Wissenschaftsgeschichte* 34, no. 3 (2011): 242–262.

37. Christine Kanz, "Max Beckmann's Revenants and Ernst Jünger's Drones: Vision and Coolness in the Interwar Period," in *Visualizing War: Images, Emotions, Communities*, ed. Anders Engberg-Pedersen and Kathrin Maurer (New York: Routledge, 2018), 51.

38. Hans Blumenberg, *Der Mann vom Mond: Über Ernst Jünger* (Frankfurt: Suhrkamp, 2007).

39. Jünger, *The Glass Bees*, 7.

40. Jünger, *The Glass Bees*, 67.

41. Jünger, *The Glass Bees*, 93.

42. Jünger, *The Glass Bees*, 134.

43. Jünger, *The Glass Bees*, 107.

44. Jünger, *The Glass Bees*, 81.

45. See Niels Werber, "Jüngers Bienen," *Zeitschrift für deutsche Philologie* 130 (2011): 245–260.

46. In this respect, I take a different approach to Jünger's glass bees than Graae's military reading of the novel: Graae, "Swarm of Steel."

47. Devin Fore, "The Entomic Age," *Grey Room* (2008): 26–55.

48. Helmut Lethen, *Verhaltenslehren der Kälte: Lebensversuche zwischen den Kriegen* (Frankfurt: Suhrkamp, 1994).

49. Kathrin Maurer, "Det Farlige Øjeblik: Ernst Jüngers Fotobøger og hans Teori om Fotografiet," in *Soldat, Arbejder, Anark: Ernst Jüngers Forfatterskab*, ed. Adam Paulsen (Copenhagen: Museum Tusculanum, 2017), 155–176.

50. Jünger, *The Glass Bees*, 94.

51. Jünger, *The Glass Bees*, 97.

52. Gilles Deleuze and Félix Guattari, *A Thousand Plateaus: Capitalism and Schizophrenia* (Minneapolis: University of Minnesota Press, 1987), 30, first published 1980 in French.

53. Michael Hardt and Antonio Negri, *Multitude: War and Democracy in the Age of Empire* (New York: Penguin, 2005)

54. Eugene Thacker, "Networks, Swarms, and Multitudes," *Ctheory.net*, May 18, 2004, https://journals.uvic.ca/index.php/ctheory/article/view/14541/5388.

55. Thacker, "Networks, Swarms, and Multitudes."

56. Thacker, "Networks, Swarms, and Multitudes."

57. Thacker, "Networks, Swarms, and Multitudes."

58. Jünger, *The Glass Bees*, 94.

59. Jünger, *The Glass Bees*, 94.

60. Jünger, *The Glass Bees*, 94.

61. Jünger, *The Glass Bees*, 56.

62. Jünger, *The Glass Bees*, 90.

63. Jünger, *The Glass Bees*, 93.

64. Jünger, *The Glass Bees*, 42.

65. Jünger, *The Glass Bees*, 112.

66. Jünger, *The Glass Bees*, 51.

67. Jünger, *The Glass Bees*, 72.

68. Jünger, *The Glass Bees*, 100.

69. Jünger, *The Glass Bees*, 119.

70. Christian Benne, "Acid Jünger: Tilnærmelser," in Paulsen, *Soldat, Arbejder, Anark*, 359–377.

71. Jünger, *The Glass Bees*, 144.

72. Fore, "The Entomic Age." Fore reads the loss of organs (ears) as a farewell to the anthropomorphic vision of technology promoted by Ernst Kapp. Kapp's idea of "organ projection" suggests that technology is based on the model of human organs (e.g., hands and hammers). However, this notion no longer works: the ears (like the bees) de-anthropomorphize technology, literally departmentalizing the body into pieces. I partly agree with this reading, but I would highlight that Kapp's organ theory can also be read as bringing together the realms of humans and technology, bridging the ontological rift between the two, and that this idea is promoted in *The Glass Bees*. I further discuss Kapp in the Conclusion to this volume.

73. Jünger, *The Glass Bees*, 142.

74. See Graae, "Swarm of Steel."

75. Christine Kanz, *Maternale Moderne: Männliche Gebärphantasien zwischen Kultur und Wissenschaft (1890–1933)* (Munich: Fink, 2009).

76. Jünger, *The Glass Bees*, 144.

77. Marcela Suarez Estrada, "Hacking the Drone: Counter-Culture Interventions to Fight Violence in Latin America," in Emmert et al., *Game of Drones*, 154–163.

78. Austin Choi-Fitzpatrick, *The Good Drone: How Social Movements Democratize Surveillance* (Cambridge, MA: MIT Press, 2020). See also Dennis Zuev and Gary Bratchford, "The Citizen Drone: Protest, Sousveillance and Droneviewing," *Visual Studies* 35, no. 5 (2020): 442–456.

79. William E. Connolly, *Facing the Planetary: Entangled Humanism and the Politics of Swarming* (Durham, NC: Duke University Press, 2017).

80. Connolly, *Facing the Planetary*, 124.

81. Connolly, *Facing the Planetary*, 125.

82. Connolly, *Facing the Planetary*, 138.

83. Connolly, *Facing the Planetary*, 132.

84. Samira Shackle, "The Mystery of the Gatwick Drone," *The Guardian*, December 1, 2020, https://www.theguardian.com/uk-news/2020/dec/01/the-mystery-of-the-gatwick -drone.

85. Choi-Fitzpatrick, *The Good Drone*, 155.

86. Simon Calder, "Extinction Rebellion Postpones Heathrow Flight Disruption Plans for the Summer," *Independent*, June 16, 2019, https://www.independent.co.uk /travel/news-and-advice/extinction-rebellion-heathrow-drones-protest-environment -terminals-flight-delays-cancelled-a8961026.html.

87. "About Us," Extinction Rebellion, accessed October 6, 2021, https://extinction rebellion.uk/the-truth/about-us/.

88. Choi-Fitzpatrick, *The Good Drone*, 141–78.

89. Nick Estes, *Our History Is the Future: Standing Rock versus the Dakota Access Pipeline, and the Long Tradition of Indigenous Resistance* (London: Verso, 2019).

90. J. D. Schnepf, "Unsettling Aerial Surveillance: Surveillance Studies after Standing Rock," *Surveillance & Society* 17, no. 5 (2019): 750; Anna Feigenbaum and Anja Kanngieser, "For a Politics of Atmospheric Governance," *Dialogues in Human Geography* 5, no. 1 (2015): 80–84.

91. *Eyes in the Sky*, directed by Frédérick A. Belzile, 2017, 0:11–0:21 mins.

92. Anthony McCosker, "Drone Media: Unruly Systems, Radical Empiricism, and Camera Consciousness," *Culture Machine* 16 (2015), https://culturemachine.net /drone-culture/drone-media/.

93. Sarah Tuck, "Drone Vision and Protest," *Photographies* 11, nos. 2–3 (2018): 169–175; Sarah Tuck, "Drone Alliances," in *Fragmentation of the Photographic Image in the Digital Age*, ed. Daniel Rubenstein (New York: Routledge, 2020), 73–80; Jessica L. Horton, "Drones and Snakes," *Art in America* 105, no. 9 (2017): 104–9.

94. Schnepf, "Unsettling Aerial Surveillance."

95. McCosker, "Drone Media," 2.

96. Steven Connor, "The Menagerie of the Senses," *The Senses and Society* 1, no. 1 (2006): 12.

97. "The military naming of drones—yielding various insectile names such as Black Hornet, Killer-bee, Wasp, Mantis, and Gnat—refers metaphorically to their visual and sonic appearance, which is encompassed by the very etymology of the word 'drone', originally denoting both a male honeybee, and the deep, machinic buzzing sound it makes." Graae, "Swarm of Steel," 150.

98. Claudette Lauzon, "Stranger Things: A Techno-Bestiary of Drones in Art and War," in Graae and Maurer, *Drone Imaginaries*, 180–202.

Chapter 7

1. Alex Williams, "The Drones Were Ready for This Moment," *New York Times*, May 23, 2020, https://www.nytimes.com/2020/05/23/style/drones-coronavirus.html.

2. Michael Richardson, "Pandemic Drones," The Conversation, March 31, 2020, https://theconversation.com/pandemic-drones-useful-for-enforcing-social-distancing-or-for-creating-a-police-state-134667.

3. Williams, "The Drones Were Ready."

4. Westport Police Department, "Westport Police Department Testing New Drone Technology 'Flatten the Curve Pilot Program,'" press release, April 21, 2020.

5. Westport Police Department, "Testing New Drone Technology."

6. Meghan Holden, "Statement Regarding Westport Drone Covid-19 Pilot Program," ACLU, April 22, 2020, https://www.acluct.org/en/press-releases/statement-regarding-westport-drone-covid-19-pilot-program.

7. Rob Kitchin, "Civil Liberties *or* Public Health, or Civil Liberties *and* Public Health? Using Surveillance Technologies to Tackle the Spread of COVID-19," *Space and Polity* 24, no. 3 (2020): 362–381; Rob Kitchin, "Using Digital Technologies to Tackle the Spread of the Coronavirus: Panacea or Folly?" Programmable City Working Paper 44, Maynooth University, Maynooth, April 2020, http://progcity.maynoothuniversity.ie/.

8. Michael Richardson, "Drone Cultures: Encounters with Everyday Militarisms," *Continuum* 34, no. 6 (2020): 858–869.

9. Siegfried Kracauer, "The Mass Ornament," in *The Mass Ornament: Weimar Essays* (Cambridge, MA: Harvard University Press, 1995), 75.

10. *Rotor DR1*, directed by Chad Kapper, Cinema Libre Studios, 2015, 99 mins.

11. Rachael D'Armore, "China Deploys Drones to Patrol Its Cities amid Coronavirus Outbreak," *Global News*, February 11, 2020, https://globalnews.ca/video/6535304/china-deploys-drones-to-patrol-its-cities-amid-coronavirus-outbreak.

12. Chas Danner, "Watch Drones Scold Civilians for Not Wearing Masks," *Intelligencer*, January 31, 2020, https://nymag.com/intelligencer/2020/01/coronavirus-watch-drones-scold-maskless-civilians-in-china.html.

13. Zoe Tidman, "Coronavirus: Italian Mayor Plans on Using Drones to Send People Back Home during Lockdown," *Euronews*, March 26, 2020, https://www.euronews.com/2020/03/26/watch-italian-mayor-uses-drone-to-scream-at-locals-to-stay-indoors-amid-coronavirus-crisis.

14. Michel Chion, *Audio-Vision: Sound on Screen* (New York: Columbia University Press, 1999).

15. Chion, *Audio-Vision*, 24.

16. Chion, *Audio-Vision*, 24.

17. Howard P. Segal, *Technological Utopianism in American Culture* (Syracuse: Syracuse University Press, 2005)

18. Maria Enrica Virgillito, "Rise of the Robots: Technology and the Threat of a Jobless Future," *Labor History* 58, no. 2 (2017): 240–242.

19. Christopher J. Coyne and Yuliya Yatsyshina, "Pandemic Police States," GMU Working Paper in Economics 20–25, May 11, 2020, https://ssrn.com/abstract=3598643.

20. Faine Greenwood, "The Dawn of the Shout Drone," *Slate*, April 16, 2020, https://slate.com/technology/2020/04/coronavirus-shout-drone-police-surveillance.html.

21. Carla Babb and Hong Xie, "US Military Still Buying Chinese-Made Drones Despite Spying Concerns," *Voa News*, September 17, 2019, https://www.voanews.com/usa/us-military-still-buying-chinese-made-drones-despite-spying-concerns.

22. On the Obama administration's "war on Ebola," see Kristin Bergtora Sandvik and Maria Gabrielsen Jumbert, "Humanitarian Drones: An Inventory," *Revue Internationale et Stratégique* 98, no. 2 (2015): 139–146.

23. Kristin Bergtora Sandvik, "African Drone Stories," *Behemoth: A Journal on Civilisation* 8, no. 2 (2015): 81.

24. Sandvik, "African Drone Stories," 81; see also Derek Gregory, "The War on Ebola," Geographical Imaginations, October 25, 2014, http://geographicalimaginations.com/2014/10/25/the-war-on-ebola.

25. Sandvik, "African Drone Stories," 78.

26. Sandvik, "African Drone Stories," 75.

27. Sandvik, "African Drone Stories," 75.

28. Kristin Bergtora Sandvik, Katja Lindskov Jacobsen, and Sean Martin McDonald, "Do No Harm: A Taxonomy of the Challenges of Humanitarian Experimentation," *International Review of the Red Cross* 99, no. 904 (2017): 319–344.

29. Ira Boudway, "Medical Drone Startup to Begin Covid Vaccine Delivery in April," *Bloomberg News*, February 4, 2021, https://www.bloomberg.com/news/articles/2021-02-04/medical-drone-startup-to-begin-covid-vaccine-delivery-in-april.

30. *Rotor DR1*, 1:34:33 mins.

31. Sarah Cascone, "This Aerial Photographer Captured Images of the Mass Burials on Hart Island," Art.net, April 20, 2020, https://news.artnet.com/art-world/nypd

-confiscates-drone-hart-island-1838187; Christopher Robbins, "NYPD Seizes Drone of Photojournalist Documenting Mass Burials on Hart Island," Gothamist, April 20, 2020, https://gothamist.com/news/nypd-seizes-drone-photojournalist-documenting -mass-burials-hart-island. I also discuss a different photographic work by Steinmetz in chapter 5.

32. Jackie Vandinther, "Drone Video Shows Inmates Digging Mass Burial Graves on New York's Hart Island," *CTV News*, April 8, 2020, https://www.ctvnews.ca/world/drone -video-shows-inmates-digging-mass-burial-graves-on-new-york-s-hart-island-1.4888134.

33. Tina Moore and Dean Balsamini, "NYPD Seizes Drone Documenting Mass Hart Island Burials amid Coronavirus," *New York Post*, April 18, 2020, https://nypost .com/2020/04/18/nypd-seizes-drone-documenting-mass-hart-island-burials-amid -coronavirus.

34. Moore and Balsamini, "NYPD Seizes Drone."

35. Grant Bollmer, "Empathy Machines," *Media International Australia* 165, no. 1 (2017): 64.

36. Robert Vischer, "On the Optical Sense of Form: A Contribution to Aesthetics," in *Empathy, Form, and Space: Problems in German Aesthetics, 1973–1893*, ed. Harry Francis Mallgrave and Eleftherios Ikonomou (Santa Monica: Getty Center for the History of Art and the Humanities, 1994), 89–123.

37. Heinrich Wölfflin, "Prolegomena to a Psychology of Architecture," in Mallgrave and Ikonomou, *Empathy, Form, and Space*, 149–187.

38. August Schmarsow, "The Essence of Architectonic Creation," in Mallgrave and Ikonomou, *Empathy, Form, and Space*, 286.

39. In addition to Vischer, Wölfflin, and Schmarsow, we might also refer to the psychologist Theodor Lipps, who expanded the idea of *Einfühlung* into human rela- tions and developed a form of aesthetic philosophy. For a discussion of aesthetic empathy, see Juliet Koss, "On the Limits of Empathy," *Art Bulletin* 88, no. 1 (2006): 139–157.

40. Today there is a wide debate about empathy and neuroscience, mirror neurons, and aesthetics. The discussion focuses on the body schema and embodied simula- tion. See Shaun Gallagher, "Aesthetics and Kinaesthetics," in *Sehen und Handeln*, vol. 1, ed. Horst Bredekamp and John M. Krois (Berlin: Akademie, 2012), 99–113.

41. Aby Warburg, *Der Bilderatlas Mnemosyne* (Berlin: Akademie, 2008).

42. Georges-Didi Hubermann, *Das Nachleben der Bilder: Kunstgeschichte und Phan- tomzeit nach Aby Warburg* (Frankfurt: Suhrkamp, 2010).

43. Antonio Somaini, "Machine Vision in Pandemic Times," in *Pandemic Media: Preliminary Notes Toward an Inventory*, ed. Philipp Dominik Keidl et al. (Lüneburg: Meson Press, 2020), 152.

44. Harun Farocki, "Phantom Images," trans. Brian Poole, *Public* 29 (2004): 12–24.

45. My research is based on the following work, although I offer a different interpretation of empathy and the drone: Somaini, "Machine Vision in Pandemic Times"; Caren Kaplan and Patricia A. Zimmermann, "Coronavirus Drone Genres: Spectacles of Distance and Melancholia," *Film Quarterly*, April 30, 2020, https://filmquarterly .org/2020/04/30/coronavirus-drone-genres-spectacles-of-distance-and-melancholia/.

46. *Il Silenzio di Roma*, directed by Luigi di Palumbo, Individo.it, April 2020, 2:31 mins., https://www.youtube.com/watch?v=pTHIXUzVGu8.

47. Shaun Gallagher and Jonathan Cole discuss the kinetic aspect of the body schema and its relation to empathy: Shaun Gallagher and Jonathan Cole, "Body Schema and Body Image in a Deafferented Subject," *Journal of Mind and Behavior* 16, no. 4 (1995): 369–390.

48. John Johnston, "Machinic Vision," *Critical Inquiry* 26 (1999): 27–48.

49. Walter Benjamin, *Berlin Childhood around 1900* (Cambridge, MA: The Belknap Press of Harvard University Press, 2006), 53–54.

50. Ada Ackerman, "Covid-Dronism: Pandemic Visions from Above," in Keidl et al., *Pandemic Media*, 168.

51. Deborah C. Beidel et al., "Trauma Management Therapy with Virtual-Reality Augmented Exposure Therapy for Combat-Related PTSD: A Randomized Controlled Trial," *Journal of Anxiety Disorders* 61 (2019): 64–74.

52. Teresa Castro, "Of Drones and the Environmental Crisis in the Year of 2020," in Keidl et al., *Pandemic Media*, 85.

53. "While the emphatic and spectacular quality conveyed by the totalizing and sweeping eye of the drone invites one to marvel at the beauty of the sites viewed, at the 'purity' of their design and architecture, since almost no human presence is obstructing the view anymore and thanks to a significant decrease in pollution, it is precisely this beauty that appears not only as paradoxical but also as problematic. It is conditioned by a radical removal of human elements, a beauty fostering a feeling of a terrifying sublime among the viewer, according to Kant's definition of a feeling of amazement mingled with dread." Ackerman, "Covid-Dronism," 169.

54. Jean-François Lyotard, *Lessons on the Analytic of the Sublime: Kant's Critique of Judgment* (Stanford: Stanford University Press, 1994).

55. Kaplan and Zimmermann, "Coronavirus Drone Genres."

56. Jan Mieszkowski, "The Drone of Data," in *Drone Imaginaries: The Remote Power of Vision*, ed. Andreas Immanuel Graae and Kathrin Maurer (Manchester: Manchester University Press, 2021), 55–73; Svea Braeunert, "Shifting the Pattern: Lateral Thinking and Machine Vision," *The Senses and Society* 15, no. 3 (2020): 259–271. My

argument is indebted to these researchers' discussions of the connection between drones and Kracauer's theory of the mass ornament. In contrast to Braeunert, however, I interpret dividualization as an opportunity to formulate new forms of communities and models of subjectivity.

57. Kracauer, "The Mass Ornament," 76.

58. Kracauer, "The Mass Ornament," 77.

59. Kracauer, "The Mass Ornament," 77.

60. On the aesthetics of the ornament and the drone, see Christoph Asendorf's discussion of Kazimir Malevich, the aerial view of flattening, and avant-garde aesthetics: Christoph Asendorf, "Bewegliche Fluchpunkte: Der Blick von oben und die moderne Raumanschauung," in *Iconic Worlds: Neue Bilderwelten und Wissensräume,* ed. Christa Maar and Hubert Burda (Cologne: DuMont, 2006), 11–16. See also Mieszkwoski, "The Drone of Data."

61. Kracauer, "The Mass Ornament," 78.

62. See Braeunert, "Shifting the Pattern," 260.

63. Gilles Deleuze, "Postscript on the Societies of Control," *October* 59 (1992): 4.

64. Gerald Raunig, *Dividuum: Machinic Capitalism and Molecular Revolution* (South Pasadena: Semiotext(e), 2016).

65. Kracauer, "The Mass Ornament," 78.

66. Some critics have taken Kracauer's critique of capitalism even further and pointed out the connections between the mass ornament and the aesthetics of fascism, evinced in Leni Riefenstahl's film *Triumph of the Will.* See Karsten Witte, Barbara Correll, and Jack Zipes, "Introduction to Siegfried Kracauer's 'The Mass Ornament,'" *New German Critique* 5, no. 1 (1975): 61.

67. Kracauer, "The Mass Ornament," 86.

68. Kracauer, "The Mass Ornament," 86.

69. Kracauer, "The Mass Ornament," 75.

70. Jay David Bolter and Richard Grusin, *Remediation: Understanding New Media* (Cambridge, MA: MIT Press, 2000), 256.

71. Yvonne Zimmermann, "Videoconferencing and the Uncanny Encounter with Oneself: Self-Reflexivity as Self-Monitoring 2.0," in Keidl et al., *Pandemic Media,* 99–103.

72. Kracauer, "The Mass Ornament," 76.

Conclusion

1. The drone collects these memories during public festivals and symposia, and they are presented in chapter publications. See the project website, accompanying chapters, and documentation: Emilio Vavarella, accessed June 2020, https://emiliovavarella .com/.

2. Pierre Nora and Lawrence D. Kritzman, eds., *Realms of Memory: Rethinking the French Past*, vols. 1–3 (New York: Columbia University Press, 1996).

3. Walter Benjamin, "Theses on the Philosophy of History," in *Illuminations: Essays and Reflections*, ed. Hannah Arendt (New York: Schocken, 1968), 253–264.

4. Emilio Vavarella, "Interview with the Drone: Experimenting with Post-Anthropocentric Art Practice," *Digital Creativity* 27, no. 1 (2016): 76.

5. For a discussion of this "premodern" future essentialism in light of technological developments, see Kasper Schiølin, "Revolutionary Dreams: Future Essentialism and the Sociotechnical Imaginary of the Fourth Industrial Revolution in Denmark," *Social Studies of Science* 50, no. 4 (2020): 542–566.

6. Louise Amoore, "Algorithmic War: Everyday Geographies of the War on Terror," *Antipode* 41, no. 1 (2009): 49–69.

7. Derek Gregory, "From a View to a Kill: Drones and Late Modern War," *Theory, Culture & Society* 28, nos. 7–8 (2014): 192.

8. See e.g., Grégoire Chamayou, *A Theory of the Drone* (New York: New Press, 2015).

9. Brian Massumi, *Ontopower: War, Powers, and the State of Perception* (Durham, NC: Duke University Press, 2015), 14. I owe this reference to Massumi and the drone to Andreas Immanuel Graae, "The Cruel Drone: Imagining Drone Warfare in Art, Culture, and Politics" (PhD diss., University of Southern Denmark, 2017), 32.

10. Paul Henry Cureton, *Drone Futures: UAS in Landscape and Urban Design* (London: Routledge, 2021), 198–199.

11. On the notion of the collective singular as a basis for shared (e.g., national) identities, see "Geschichte: Historie," in *Geschichtliche Grundbegriffe: Historisches Lexikon zur poltisch-sozialen Sprache in Deutschland*, vol. 2, ed. Otto Brunner, Werner Conze, and Reinhart Koselleck (Stuttgart: Klett, 1972), 647–652.

12. Walter Benjamin, *The Work of Art in the Age of Its Technological Reproducibility, and Other Writings on Media*, ed. Michael W. Jennings, Brigid Doherty, and Thomas Y. Levin (Cambridge, MA: The Belknap Press of Harvard University Press, 2008), 42.

13. Ernst Kapp, *Elements of a Philosophy of Technology: On the Evolutionary History of Culture* (Minneapolis: University of Minnesota Press, 2018).

14. Yuk Hui, "Machine and Ecology," *Angelaki* 25, no. 4 (2020): 54–66.

15. Jeffrey West Kirkwood and Leif Weatherby, "Operations of Culture: Ernst Kapp's Philosophy of Technology," *Grey Room* (2018): 6–15.

16. Melvin Kranzberg, "Technology and History: Kranzberg's Laws," *Technology and Culture* 27, no. 3 (1986): 544–560.

Afterword

1. Martin Heidegger, *The Question Concerning Technology and Other Essays*, ed. and trans. William Lovitt (New York: Garland, 1977), 4.

2. Heidegger, *The Question Concerning Technology*, 13.

3. Gerhard Banse and Armin Grunwald, *Technik und Kultur: Bedingungs- und Beein-flussungsverhältnisse* (Karlsruhe: KIT Scientific, 2010).

4. Gernot Böhme, *Invasive Technification: Critical Essays in the Philosophy of Technology* (London: Bloomsbury, 2012).

Bibliography

Ackerman, Ada. "Covid-Dronism: Pandemic Visions from Above." In *Pandemic Media: Preliminary Notes toward an Inventory*, edited by Philipp Dominik Keidl, Laliv Melamed, Vinzenz Hediger, and Antonio Somaini, 167–172. Lüneburg: Meson Press, 2020.

Acosta, Ignacio. *Copper Geographies*. Barcelona: Editorial RM, 2018.

Adey, Peter. "Making the Drone Strange: The Politics, Aesthetics and Surrealism of Levitation." *Geographica Helvetica* 71, no. 4 (2016): 319–329.

Adey, Peter, Mark Whitehead, and Alison J. Williams. "Introduction: Air-Target Distance, Reach and the Politics of Verticality." *Theory, Culture & Society* 27, no. 6 (2012): 173–187.

Agostinho, Daniela, Kathrin Maurer, and Kristin Veel, eds. "The Sensorial Experience of the Drone." Themed issue, *The Senses and Society* 15, no. 3 (2020).

Alkouz, Balsam, Athman Bouguettaya, and Sajib Mistry. "Swarm-Based Drone-as-a-Service (SDAAS) for Delivery." In *2020 IEEE International Conference on Web Services Proceedings*, edited by Lisa O'Connor, 441–448. Los Alamitos: Institute of Electrical and Electronic Engineers, 2020.

Alpers, Svetlana. *The Art of Describing: Dutch Art in the Seventeenth Century*. Chicago: University of Chicago Press, 1984.

Alvarez-Taboada, Flor, Claudio Paredes, and Julia Julián-Pelaz. "Mapping of the Invasive Species *Hakea sericea* Using Unmanned Aerial Vehicle (UAV) and WorldView-2 Imagery and an Object-Oriented Approach." *Remote Sensing* 9, no. 9 (2017): 1–17.

Amelang, Katrin. "Zur Sinnlichkeit von Algorithmen und ihrer Erforschbarkeit." In *Kulturen der Sinne: Zugänge zur Sensualität der sozialen Welt*, edited by Karl Braun, Claus-Marco Dieterich, Thomas Hengartner, and Bernhard Tschofen, 358–367. Würzburg: Königshausen Neumann, 2017.

Amoore, Louise. "Algorithmic War: Everyday Geographies of the War on Terror." *Antipode* 41, no. 1 (2009): 49–69.

Amoroso, Nadia. *The Exposed City: Mapping the Urban Invisibles*. New York: Routledge, 2010.

Anand, Abhishek, Peng Wei, Nirmal Kumar Gali, Li Sun, Fenhuan Yang, Dane Westerdahl, Qing Zhang. "Protocol Development for Real-Time Ship Fuel Sulfur Content Determination Using Drone-Based Plume Sniffing Microsensor System." *Science of the Total Environment* 744 (2020): 140885.

Anderson, Benedict. *Imagined Communities: Reflections on the Origin and Spread of Nationalism*. London: Verso, 2006.

Andreas, Michael. "Flächen/Rastern: Zur Bildlichkeit der Drohne." *Behemoth: A Journal on Civilisation* 8, no. 2 (2015): 108–127.

Andrejevic, Mark. "Theorizing Drones and Droning Theory." In *Drones and Unmanned Aerial Systems: Legal and Social Implications for Security and Surveillance*, edited by Aleš Završnik, 21–43. Cham: Springer, 2016.

Anker, Peder. *Imperial Ecology: Environmental Order in the British Empire, 1895–1945*. Cambridge, MA: Harvard University Press, 2001.

Armstrong, Keith. "Embodying a Future for the Future: Creative Robotics and Ecosophical Praxis." *Fibreculture Journal* 28 (2016): 1–20.

Armstrong, Keith. *Future-Future?* Mount Gambier: Riddoch Art Gallery, 2017. https://www.vrystaatkunstefees.co.za/wp-content/uploads/2020/05/Future-Future-2017.pdf.

Armstrong, Keith, and Tania Leimbach. "Art-Eco-Science: Field Collaborations." *Antennae: The Journal of Nature in Visual Culture* 48 (2019): 108–127.

Arquilla, John, and David Ronfeldt. *Swarming and the Future of Conflict*. Santa Monica: Rand, 2000.

Asaro, Peter. "The Labor of Surveillance and Bureaucratized Killing: New Subjectivities of Military Drone Operators." *Social Semiotics* 23, no. 2 (2013): 1–29.

Ascott, Roy. "Telenoia." In *Telematic Embrace: Visionary Theories of Art, Technology, and Consciousness*, edited by Edward A. Shanken, 257–275. Berkeley: University of California Press, 2003.

Asendorf, Christoph. "Bewegliche Fluchpunkte: Der Blick von oben und die moderne Raumanschauung." In *Iconic Worlds: Neue Bilderwelten und Wissensräume*, edited by Christa Maar and Hubert Burda, 11–16. Cologne: DuMont, 2006.

Augustine. *The Confessions of St. Augustine*. Translated by Rex Warner. New York: Penguin Putnam, 2001.

Azoulay, Ariella. *Civil Imagination: A Political Ontology of Photography*. London: Verso, 2012.

Babb, Carla, and Hong Xie. "US Military Still Buying Chinese-Made Drones Despite Spying Concerns." Voa News, September 17, 2019. https://www.voanews.com/usa /us-military-still-buying-chinese-made-drones-despite-spying-concerns.

Bailly, Christian. *Automata: The Golden Age, 1848–1914*. London: Robert Hale, 2003.

Baldwin, Thomas. *Airopaidia*. London: J. Fletcher, 1786.

Banse, Gerhard, and Armin Grunwald. *Technik und Kultur: Bedingungs- und Beeinflussungsverhältnisse*. Karlsruhe: KIT Scientific, 2010.

Basso, Bruno, James Dobrowski, and Channing McKay. "From the Dust Bowl to Drones to Big Data: The Next Revolution in Agriculture." *Georgetown Journal of International Affairs* 18, no. 3 (2017): 158–165.

Bauman, Zygmunt, and David Lyon. *Liquid Surveillance: A Conversation*. Hoboken: John Wiley & Sons, 2013.

Baumgarten, Alexander Gottlieb von. *Ästhetik*, edited by Dagmar Mirbach. Hamburg: Felix Meiner, 2007. First published 1750.

Baumgarten, Alexander Gottlieb von. *Philosophische Briefe des Aletheophilus*. Frankfurt/Leipzig, 1741.

Beesley, David. "Drone Panic! On Representations of the Personal Drone by Australian Mainstream Media." *Screen Thought* 1, no. 1 (2018): 1–18.

Begely, Adam. *The Great Nadar: The Man behind the Camera*. New York: Penguin Random House, 2017.

Beidel, Deborah C., Christopher Frueh, Sandra M. Neer, Clint A. Bowers, Benjamin Trachik, Thomas W. Uhde, and Anouk Grubaugh. "Trauma Management Therapy with Virtual-Reality Augmented Exposure Therapy for Combat-Related PTSD: A Randomized Controlled Trial." *Journal of Anxiety Disorders* 61 (2019): 64–74.

Belhumeur, Peter N., Joao P. Hespanha, and David J. Kriegman. "Eigenfaces vs. Fisherfaces: Recognition Using Class Specific Linear Projection." *IEEE Transactions on Pattern Analysis and Machine Intelligence* 19, no. 7 (1997): 711–720.

Benjamin, Walter. *Berlin Childhood around 1900*. Translated by Howard Eiland. Cambridge, MA: The Belknap Press of Harvard University Press, 2006.

Benjamin, Walter. *Illuminations: Essays and Reflections*. Edited by Hannah Arendt. Translated by Harry Zohn. New York: Schocken, 1968.

Benjamin, Walter. "On Some Motifs in Baudelaire." In *Illuminations: Essays and Reflections*, edited by Hannah Arendt, translated by Harry Zohn, 155–200. New York: Schocken, 1968.

Benjamin, Walter. "Theses on the Philosophy of History." In *Illuminations: Essays and Reflections*, edited by Hannah Arendt, translated by Harry Zohn, 253–264. New York: Schocken, 1968.

Benjamin, Walter. "The Work of Art in the Age of Mechanical Reproduction." In *Illuminations: Essays and Reflections*, edited by Hannah Arendt, translated by Harry Zohn, 217–251. New York: Schocken, 1968.

Benjamin, Walter. *The Work of Art in the Age of Its Technological Reproducibility, and Other Writings on Media*. Edited by Michael W. Jennings, Brigid Doherty, and Thomas Y. Levin. Translated by Edmund Jephcott, Rodney Livingstone, and Howard Eiland. Cambridge, MA: The Belknap Press of Harvard University Press, 2008.

Benne, Christian. "Acid Jünger: Tilnærmelser." Translated by Adam Paulsen. In *Soldat, Arbejder, Anark: Ernst Jüngers Forfatterskab*, edited by Adam Paulsen, 359–377. Copenhagen: Museum Tusculanum, 2017.

Berkowitz, Roger. "Drones and the Question of the Human." *Ethics & International Affairs* 28, no. 2 (2014): 159–169.

Berlant, Lauren. "The Commons: Infrastructures for Troubling Times." *Environment and Planning: Society and Space* 34, no. 3 (2016): 393–419.

Bettridge, Thom. "The Fabric of Existence: A Studio Visit with Denim Painter Korakrit Arunanondchai in New York." Ssense.com, accessed October 2, 2021. https://www.ssense.com/en-us/editorial/fashion/the-fabric-of-existence.

Blum, Hester. "Terraqueous Planet: The Case for Oceanic Studies." In *The Planetary Turn: Relationality and Geoaesthetics in the Twenty-First Century*, edited by Amy J. Elias and Christian Moraru, 25–36. Evanston: Northwestern University Press, 2015.

Blumenberg, Hans. *Der Mann vom Mond: Über Ernst Jünger*. Frankfurt: Suhrkamp, 2007.

Boehm, Gottfried. "Zwischen Auge und Hand: Bilder als Instrumente der Erkenntnis." In *Mit dem Auge Denken: Strategien der Sichtbarmachung in wissenschaftlichen und virtuellen Welten*, edited by Bettina Heintz and Jorg Huber, 43–54. Vienna: Springer, 2001.

Böhme, Gernot. *Invasive Technification: Critical Essays in the Philosophy of Technology*. London: Bloomsbury, 2012.

Bollmer, Grant. "Empathy Machines." *Media International Australia* 165, no. 1 (2017): 63–76.

Bollmer, Grant, and Katherine Guinness. "Do You Really Want to Live Forever? Animism, Death, and the Trouble of Digital Images." *Cultural Studies Review* 24, no. 2 (2018): 79–96.

Bolman, Brad. "Provocation: A Prairie Drone Companion." *Culture Machine* 16 (2015): 1–6.

Bolman, Brad. "A Revolution in Agricultural Affairs: Dronoculture, Precision, Capital." In *The Good Drone*, edited by Kristin Bergtora Sandvik and Maria Gabrielsen Jumbert, 139–162. New York: Routledge, 2016.

Bolter, Jay David, and Richard Grusin. *Remediation: Understanding New Media*. Cambridge, MA: MIT Press, 2000.

Bonmatin, Jean-Marc, Mark Brown, Robert Paxton, Michael Kuhlmann, Dave Goulson, Axel Decourtye, and Pat Willmer. "Entomology: The Bee-All and End-All." *Nature* 521 (2015): 57–59.

Bonneuil, Christophe, Jean-Baptiste Fressoz, and David Fernbach, eds. *The Shock of the Anthropocene: The Earth, History and Us*. London: Verso, 2016.

Boudway, Ira. "Medical Drone Startup to Begin Covid Vaccine Delivery in April." *Bloomberg News*, February 4, 2021. https://www.bloomberg.com/news/articles/2021 -02-04/medical-drone-startup-to-begin-covid-vaccine-delivery-in-april.

Bousquet, Antoine. *The Eye of War: Military Perception from the Telescope to the Drone*. Minneapolis: University of Minnesota Press, 2018.

Bracken-Roche, Ciara. "Domestic Drones: The Politics of Verticality and the Surveillance Industrial Complex." *Geographica Helvetica* 71, no. 3 (2016): 167–172.

Braeunert, Svea. "Disappearing, Appearing, and Reappearing: Imaging the Human Body in Drone Warfare." In *Drone Imaginaries: The Remote Power of Vision*, edited by Andreas Immanuel Graae and Kathrin Maurer, 91–109. Manchester: Manchester University Press, 2021.

Braeunert, Svea. "Shifting the Pattern: Lateral Thinking and Machine Vision." *The Senses and Society* 15, no. 3 (2020): 259–271.

Braeunert, Svea. "To See without Being Seen: Contemporary Art and Drone Warfare." In *To See without Being Seen: Contemporary Art and Drone Warfare*, edited by Svea Braeunert and Meredith Malone, 11–25. Chicago: University of Chicago Press, 2016.

Braeunert, Svea, and Meredith Malone, eds. *To See without Being Seen: Contemporary Art and Drone Warfare*. Chicago: University of Chicago Press, 2016.

Braidotti, Rosi. *Metamorphoses: Towards a Materialist Theory of Becoming*. Hoboken: John Wiley & Sons, 2013.

Braidotti, Rosi. *The Posthuman*. Cambridge: Polity, 2013.

Braidotti, Rosi. *Posthuman Knowledge*. Cambridge: Polity, 2019.

Brant, George. *Grounded*. London: Oberon Books, 2013.

Bratton, Benjamin. "Further Trace Effects of the Post-Anthropocene." *Architectural Design* 89, no. 1 (2019): 14–21.

Bredekamp, Horst. *Theorie des Bildakts*. Frankfurt: Suhrkamp, 2010.

Browne, Simone. *Dark Matters: On the Surveillance of Blackness*. Durham: Duke University Press, 2015.

Brunelli, Roberto, and Tomaso Poggio. "Face Recognition: Features versus Templates." *IEEE Transactions on Pattern Analysis and Machine Intelligence* 15, no. 10 (1993): 1042–1052.

Brunner, Otto, Werner Conze, and Reinhart Koselleck, eds. *Geschichtliche Grundbegriffe: Historisches Lexikon zur poltisch-sozialen Sprache in Deutschland*. Vol. 2. Stuttgart: Klett, 1972.

Buddemeier, Heinz. *Panorama, Diorama, Photographie: Entstehung und Wirkung neuer Medien im 19. Jahrhundert*. Munich: Fink, 1977.

Bull, Michael, Paul Gilroy, David Howes, and Douglas Kahn. "Introducing Sensory Studies." *The Senses and Society* 1, no. 1 (2006): 5–7.

Buolamwini, Joy, and Timnit Gebru. "Gender Shades: Intersectional Accuracy Disparities in Commercial Gender Classification." *Proceedings of Machine Learning Research* 81 (2018): 77–91.

Burke, Edmund. *Philosophical Enquiry into the Origin of Our Ideas of the Sublime and Beautiful*. Oxford: Oxford University Press, 2015. First published 1757.

Burtynsky, Edward, Jennifer Baichwal, and Nicolas de Pencier. *Anthropocene*. Göttingen: Steidl, 2018.

Butler, Judith. *Bodies That Matter: On the Discursive Limits of Sex*. London: Routledge, 1993.

Butler, Judith. *Gender Trouble: Feminism and the Subversion of Identity*. London: Routledge, 1990.

Calder, Simon. "Extinction Rebellion Postpones Heathrow Flight Disruption Plans for the Summer." *Independent*, June 16, 2019. https://www.independent.co.uk/travel/news-and-advice/extinction-rebellion-heathrow-drones-protest-environment-terminals-flight-delays-cancelled-a8961026.html.

Cascone, Sarah. "This Aerial Photographer Captured Images of the Mass Burials on Hart Island." *Art.net*, April 20, 2020. https://news.artnet.com/art-world/nypd-confiscates-drone-hart-island-1838187.

Castro, Teresa. "Of Drones and the Environmental Crisis in the Year of 2020." In *Pandemic Media: Preliminary Notes Toward an Inventory*, edited by Philipp Dominik Keidl,

Laliv Melamed, Vinzenz Hediger, and Antonio Somaini, 81–90. Lüneburg: Meson Press, 2020.

Chamayou, Grégoire. *Drone Theory*. Translated by Janet Lloyd. London: Penguin, 2015.

Chamayou, Grégoire. *A Theory of the Drone*. New York: New Press, 2015.

Chandler, Katherine. "A Drone Manifesto: Re-forming the Partial Politics of Targeted Killing." *Catalyst: Feminism, Theory, and Technoscience* 2, no. 1 (2016): 1–23.

Chandler, Katherine. *Unmanning: How Humans, Machines and Media Perform Drone Warfare*. New Brunswick: Rutgers University Press, 2020.

Chang, Chein-I. *Hyperspectral Imaging: Techniques for Spectral Detection and Classification*. New York: Kluwer Academic/Plenum, 2003.

Chion, Michel. *Audio-Vision: Sound on Screen*. Translated by Claudia Gorbman. New York: Columbia University Press, 1999.

Choi-Fitzpatrick, Austin. *The Good Drone: How Social Movements Democratize Surveillance*. Cambridge, MA: MIT Press, 2020.

Christiansen, Steen Ledet. "Unruly Vision: Synesthetic Space: Drone Music Videos." *Senses and Society* 15, no. 3 (2020): 286–298.

Christopher, John. *Balloons at War: Gasbags, Flying Bombs, and Cold War Secrets*. Stroud: Tempus, 2004.

Cirque du Soleil. *Sparked: Behind the Technology*. YouTube, 2:39, uploaded September 22, 2014. https://www.youtube.com/watch?v=7YqUocVcyrE.

Connolly, William E. *Facing the Planetary: Entangled Humanism and the Politics of Swarming*. Durham, NC: Duke University Press, 2017.

Connor, Steven. "The Menagerie of the Senses." *The Senses and Society* 1, no. 1 (2006): 9–26.

Cosgrove, Denis, and Carmen Cosgrove. *Apollo's Eye: A Cartographic Genealogy of the Earth in the Western Imagination*. Baltimore: Johns Hopkins University Press, 2003.

Cosgrove, Denis, and William L. Fox. *Photography and Flight*. London: Reaktion, 2010.

Coyne, Christopher J., and Yuliya Yatsyshina. "Pandemic Police States." GMU Working Paper in Economics 20–25, May 11, 2020. https://ssrn.com/abstract=3598643.

Crabtree, Jason. "Gaza and Nagorno-Karabakh Were Glimpses of the Future of Conflict." *FP*, June 21, 2021. https://foreignpolicy.com/2021/06/21/gaza-nagorno-karabakh-future-conflict-drones/.

Crampton, Jeremy. "Assemblage of the Vertical: Commercial Drones and Algorithmic Life." *Geographica Helvetica* 71 (2016): 137–146.

Crandall, Jordan. "Ecologies of a Wayward Drone." In *From Above: War, Violence and Verticality*, edited by Peter Adey, Mark Whitehead, and Alison J. Williams, 263–287. London: Hurst, 2011.

Crary, Jonathan. *Techniques of the Observer: On Vision and Modernity*. Cambridge, MA: MIT Press, 1992.

Cravey, Altha, Joseph Palis, and Gabriela Valdivia. "Imagining the Future from the Margins: Cyborg Labor in Alex Rivera's *Sleep Dealer*." *GeoJournal* 80 (2015): 867–880.

Cubitt, Sean. *EcoMedia*. Amsterdam: Rodopi, 2005.

Cureton, Paul Henry. *Drone Futures: UAS in Landscape and Urban Design*. London: Routledge, 2021.

Daggett, Cara. "Drone Disorientations: How 'Unmanned' Weapons Queer the Experience of Killing in War." *International Feminist Journal of Politics* 17, no. 3 (2015): 361–379.

Danner, Chas. "Watch Drones Scold Civilians for Not Wearing Masks." *Intelligencer*, January 31, 2020. https://nymag.com/intelligencer/2020/01/coronavirus-watch -drones-scold-maskless-civilians-in-china.html.

D'Armore, Rachael. "China Deploys Drones to Patrol Its Cities amid Coronavirus Outbreak." *Global News*, February 11, 2020. https://globalnews.ca/video/6535304 /china-deploys-drones-to-patrol-its-cities-amid-coronavirus-outbreak.

Daston, Lorraine, and Peter Galison. "The Image of Objectivity." *Representations* 40 (1992): 81–128.

Deiss, Amely, Anja Dorn, Elena Filipovic, and Milena Mercer, eds. *Raphaela Vogel*. Cologne: Walther König, 2018.

Deleuze, Gilles. "Postscript on the Societies of Control." *October* 59 (1992): 3–7.

Deleuze, Gilles, and Félix Guattari. *A Thousand Plateaus: Capitalism and Schizophrenia*. Translated by Brian Massumi. Minneapolis: University of Minnesota Press, 1987. First published 1980 in French.

Delmont, Matt. "Drone Encounters: Noor Behram, Omer Fast, and Visual Critiques of Drone Warfare." *American Quarterly* 65, no. 1 (2013): 193–202.

Droid Guy. "7 Best Drones with Facial Recognition." The Droid Guy, May 2022. https://thedroidguy.com/7-best-drones-with-facial-recognition-1077684.

Eckert, Alfred. "Zur Geschichte der Ballonfahrt." In *Leichter als Luft: Zur Geschichte der Ballonfahrt*, edited by Bernard Korzus and Gisela Noehles, 13–128. Münster: Westfälisches Landesmuseum für Kunst und Kulturgeschichte, 1978.

Eder, Jens, and Charlotte Klonk, eds. *Image Operations: Visual Media and Political Conflict*. Manchester: Manchester Press, 2016.

Ehemann, Rose. "Selfies and Dronies: Zu Chancen und Gefährdungen der Dynamisierung des Selbst in virtuellen Welten." *Journal für Psychoanalyse* 59 (2018): 60–73.

Elden, Stuart. "Secure the Volume: Vertical Geopolitics and the Depth of Power." *Political Geography* 1, no. 34 (2013): 35–51.

Elias, Amy J., and Christian Moraru, eds. *The Planetary Turn: Relationality and Geoaesthetics in the Twenty-First Century.* Evanston: Northwestern University Press, 2015.

Elster, Jon. "Exploring Exploitation." *Journal of Peace Research* 15, no. 1 (1978): 3–17.

Engberg-Pedersen, Anders. "Technologies of Experience: Harun Farocki's Serious Games and Military Aesthetics." *boundary 2* 44, no. 4 (2017): 155–178.

Engberg-Pedersen, Anders, and Kathrin Maurer, eds. *Visualizing War: Images, Emotions, Communities.* New York: Routledge, 2018.

Estes, Nick. *Our History Is the Future: Standing Rock versus the Dakota Access Pipeline, and the Long Tradition of Indigenous Resistance.* London: Verso, 2019.

Estrada, Marcela Suarez. "Hacking the Drone: Counter-Culture Interventions to Fight Violence in Latin America." In *Game of Drones: Of Unmanned Aerial Vehicles,* edited by Claudia Emmert, Jürgen Bleibler, Ina Neddermeyer, and Dominik Busch, 154–163. Berlin: Neofelis, 2020.

Farocki, Harun. "Phantom Images." Translated by Brian Poole. *Public* 29 (2004): 12–24.

Farrier, David. *Anthropocene Poetics: Deep Time, Sacrifice Zones, and Extinction.* Minneapolis: University of Minnesota Press, 2019.

Fazi, Beatrice M. "Digital Aesthetics: The Discrete and the Continuous." *Theory, Culture & Society* 36, no. 1 (2019): 3–26.

Faßler, Manfred. "Stile der Anwesenheit: Technologien, Traumgesichter, Medien." In *Wunschmaschine/Welterfindung: Eine Geschichte der Technikvisionen seit dem 18. Jahrhundert,* edited by Brigitte Felderer, 251–271. Vienna: Springer, 1996.

Feigenbaum, Anna. "From Cyborg Feminism to Drone Feminism: Remembering Women's Anti-Nuclear Activisms." *Feminist Theory* 16, no. 3 (2015): 265–288.

Feigenbaum, Anna, and Anja Kanngieser. "For a Politics of Atmospheric Governance." *Dialogues in Human Geography* 5, no. 1 (2015): 80–84.

Fernback, Jan. "Sousveillance: Communities of Resistance to the Surveillance Environment." *Telematics and Informatics* 30, no. 1 (2013): 11–21.

Fish, Adam. "Drones at the Edge of Naturecultures." *Media Fields* 14 (2019): 1–5.

Fore, Devin. "The Entomic Age." *Grey Room* (2008): 26–55.

Foucault, Michel. *Discipline and Punish: The Birth of the Prison.* Translated by Alan Sheridan. New York: Pantheon, 1977.

Franz, Nina, and Moritz Queisner. "Die Akteure verlassen die Kontrollstation: Krisen-hafte Kooperationen im bildgeführten Drohnenkrieg." In *Das Mitsein der Medien: Prekäre Koexistenzen von Menschen, Maschinen und Algorithmen*, edited by Johannes Bennke, Johanna Seifert, Martin Siegler, and Christina Terberl, 27–58. Munich: Fink, 2018.

French, Sally. "Meet the World's Top Female Drone Pilot." The Drone Girl, April 25, 2016. http://thedronegirl.com/2016/04/25/meet-zoe-the-worlds-top-female-drone -pilot/.

Frisk, Adam. "What Is Project Maven? The Pentagon AI Project Google Employees Want Out Of." *Global News*, April 5, 2018. https://globalnews.ca/news/4125382/google -pentagon-ai-project-maven/.

Fryer-Biggs, Zachary. "Inside the Pentagon's Plan to Win over Silicon Valley's AI Experts." *Wired*, December 21, 2018. https://www.wired.com/story/inside-the-penta gons-plan-to-win-over-silicon-valleys-ai-experts/.

Gabrys, Jennifer. *Digital Rubbish: A Natural History of Electronics*. Ann Arbor: Univer-sity of Michigan Press, 2011.

Gabrys, Jennifer. *Program Earth: Environmental Sensing Technology and the Making of a Computational Planet*. Minneapolis: University of Minnesota Press, 2016.

Gallagher, Shaun. "Aesthetics and Kinaesthetics." In *Sehen und Handeln*, vol. 1, edited by Horst Bredekamp and John M. Krois, 99–113. Berlin: Akademie, 2012.

Gallagher, Shaun, and Jonathan Cole. "Body Schema and Body Image in a Deaf-ferented Subject." *Journal of Mind and Behavior* 16, no. 4 (1995): 369–390.

Galloway, Alexander R., and Eugene Thacker. *The Exploit: A Theory of Networks*. Min-neapolis: University of Minnesota Press, 2013.

Garret, Bradley L., and Anthony McCosker. "Non-human Sensing: New Methodolo-gies for the Drone Assemblage." In *Refiguring Techniques in Digital Visual Research*, edited by Edgar Gómez Cruz, Shanti Sumartojo, and Sarah Pink, 13–23. London: Palgrave Macmillan, 2017.

Gates, Kelly A. *Our Biometric Future: Facial Recognition Technology and the Culture of Surveillance*. New York: NYU Press, 2011.

Gibbs, Samuel. "Google's AI Is Being Used by US Military Drone Programme." *The Guardian*, March 7, 2018. https://www.theguardian.com/technology/2018/mar/07 /google-ai-us-department-of-defense-military-drone-project-maven-tensorflow.

Girot, Christophe. "Cloudism: Towards a New Culture of Making Landscapes." In *Routledge Research Companion to Landscape Architecture*, edited by Ellen Braae and Henriette Steiner, 113–123. London: Routledge, 2018.

Girot, Christophe, Ilmar Hurkxkens, Alexandre Kapellos, James Melsom, and Pascal Werner. "Scales of Topology in Landscape Architecture." *Scales of the Earth* 4 (2011): 156–163.

Girot, Christophe, and James Melsom. "The Return of the Aviators." *Topos* 86 (2014): 102–107.

Gitelman, Lisa, ed. *Raw Data Is an Oxymoron.* Cambridge, MA: MIT Press, 2013.

Global News. "China Deploys Drones to Patrol Its Cities amid Coronavirus Outbreak." *Global News,* February 11, 2020. https://globalnews.ca/video/6535304/china -deploys-drones-to-patrol-its-cities-amid-coronavirus-outbreak.

Graae, Andreas Immanuel. "The Cruel Drone: Imagining Drone Warfare in Art, Culture, and Politics." PhD diss., University of Southern Denmark, 2017.

Graae, Andreas. "Swarming Sensations: Robo-bees and the Politics of the Swarm in *Black Mirror.*" *The Senses and Society* 15, no. 3 (2020): 329–343.

Graae, Andreas Immanuel. "Swarm of Steel: Insects, Drones and Swarming in Ernst Jünger's *The Glass Bees.*" In *Drone Imaginaries: The Remote Power of Vision,* edited by Andreas Immanuel Graae and Kathrin Maurer, 149–166. Manchester: Manchester University Press, 2021.

Graae, Andreas Immanuel, and Kathrin Maurer. "Introduction." In *Drone Imaginaries: The Remote Power of Vision,* edited by Andreas Immanuel Graae and Kathrin Maurer, 1–16. Manchester: Manchester University Press, 2021.

Graham, Kennedy, ed. *The Planetary Interest: A New Concept for the Global Age.* New Brunswick: Rutgers University Press, 1999.

Graham, Stephen. "Drone: The Robot Imperium." In *Vertical: The City from Satellites to Bunkers,* 67–94. London: Verso, 2016.

Graham, Stephen. *Vertical: The City from Satellites to Bunkers.* London: Verso, 2016.

Grau, Oliver. *Virtual Art: From Illusion to Immersion.* Cambridge, MA: MIT Press, 2003.

Grayson, Kyle, and Jocelyn Mawdsley. "Scopic Regimes and the Visual Turn in International Relations: Seeing World Politics through the Drone." *European Journal of International Relations* 25, no. 2 (2019): 431–457.

Greene, Daniel. "Drone Vision." *Surveillance & Society* 13, no. 2 (2015): 233–249.

Greenwood, Faine. "Can a Police Drone Recognize Your Face?" *Slate,* July 8, 2020. https://slate.com/technology/2020/07/police-drone-facial-recognition.html.

Greenwood, Faine. "The Dawn of the Shout Drone." *Slate,* April 16, 2020. https:// slate.com/technology/2020/04/coronavirus-shout-drone-police-surveillance.html.

Gregory, Derek. "The Everywhere War." *Geographical Journal* 177, no. 3 (2011): 238–250.

Gregory, Derek. "From a View to a Kill: Drones and Late Modern War." *Theory, Culture & Society* 28, nos. 7–8 (2014): 188–215.

Gregory, Derek. "Seeing Machines." Geographical Imaginations: Wars, Spaces and Bodies, April 15, 2014. https://geographicalimaginations.com/2014/04/15/seeing -machines/.

Gregory, Derek. "The War on Ebola." Geographical Imaginations: Wars, Spaces and Bodies, October 25, 2014. http://geographicalimaginations.com/2014/10/25/the-war -on-ebola.

Grimal, Francis, and Jae Sundaram. "Combat Drones: Hives, Swarms, and Autonomous Action?" *Journal of Conflict and Security Law* 23, no. 1 (2018): 105–135.

Gusterson, Hugh. *Drone: Remote Control Warfare*. Cambridge, MA: MIT Press, 2015.

Gynnild, Astrid. "The Robot Eyewitness: Extending Visual Journalism through Drone Surveillance." *Digital Journalism* 2, no. 3 (2014): 334–433.

Hansen, John Paulin, Alexandre Alapetite, Ian Scott MacKenzie, and Emilie Møllenbach. "The Use of Gaze to Control Drones." In *Proceedings of the Symposium on Eye Tracking Research and Application*, edited by Pernilla Qvarfordt and Dan Wintzer Hansen, 27–34. New York: The Association of Computing Machinery: 2014.

Haraway, Donna, *Manifestly Haraway*. Minneapolis: University of Minnesota Press, 2016.

Haraway, Donna. "Situated Knowledges: The Science Question in Feminism and the Privilege of Partial Perspective." *Feminist Studies* 14, no. 3 (1988): 575–599.

Hardt, Michael, and Antonio Negri. *Multitude: War and Democracy in the Age of Empire*. New York: Penguin, 2005.

Haus der Kunst. *Künstlergespräch mit Raphaela Vogel und der Philosophin Juliane Rebentisch*. YouTube, 1:14:03, uploaded July 1, 2019. https://www.youtube.com/watch ?v=7bkiHozNxOk.

Hawkins, Gay. *The Ethics of Waste: How We Relate to Rubbish*. Lanham: Rowman and Littlefield, 2005.

Heidegger, Martin. "Die Frage nach der Technik, Vorträge und Aufsätze (1910–1976)." In *Gesamtausgabe*, vol. 7, edited by Friedrich Wilhelm von Herrmann, 5–36. Frankfurt: Vittorio Klostermann, 2000.

Heidegger, Martin. *The Question Concerning Technology and Other Essays*. Edited and translated by William Lovitt. New York: Garland, 1977.

Heise, Ursula. *Sense of Place and Sense of Planet: The Environmental Imagination of the Global*. New York: Oxford University Press, 2010.

Helmholtz, Hermann von. *Die Thatsachen in der Wahrnehmung*. Berlin: August Hirschwald, 1879.

Herrmann, Goetz, and Jutta Weber. "Game of Swarms: Swarm Technologies, Control, and Autonomy in Complex Weapons Systems." In *Game of Drones: Of Unmanned Aerial Vehicles*, edited by Claudia Emmert, Jürgen Bleibler, Ina Neddermeyer, and Dominik Busch, 174–196. Berlin: Neofelis, 2020.

Hillenbrand, Tom. *Drohnenland*. Cologne: Kiepenheuer und Witsch, 2015.

Hoel, Aud Sissel. "Operative Images: Inroads to a New Paradigm of Media Theory." In *Image-Action-Space: Situating the Screen in Visual Practice*, edited by Louisa Feiersinger, Kathrin Friedrich, and Moritz Queisner, 11–28. Berlin: Walter de Gruyter, 2018.

Hoelzl, Ingrid. "Postimage." In *Posthuman Glossary*, edited by Rosi Braidotti and Maria Hlavajova, 360–362. London: Bloomsbury, 2018.

Hoelzl, Ingrid, and Remi Marie. "From Softimage to Postimage." *Leonardo* 50, no. 1 (2017): 72–73.

Holden, Meghan. "Statement Regarding Westport Drone Covid-19 Pilot Program." ACLU, April 22, 2020. https://www.acluct.org/en/press-releases/statement-regarding-westport-drone-covid-19-pilot-program.

Holland, Arthur Michel. *Eyes in the Sky: The Secret Rise of Gorgon Stare and How It Will Watch Us All*. Boston, MA: Houghton Mifflin Harcourt, 2019.

Holmqvist, Caroline. "Undoing War: War Ontologies and the Materiality of Drone Warfare." *Millennium* 41, no. 3 (2014): 535–552.

Horn, Eva. "Challenges for an Aesthetics of the Anthropocene." In *The Anthropocenic Turn: The Interplay between Disciplinary and Interdisciplinary Responses to a New Age*, edited by Gabriele Dürbeck and Philip Hüpkes, 97–111. New York: Routledge, 2020.

Horn, Eva, and Lucas Marco Gisi, eds. *Schwärme: Kollektive ohne Zentrum: Eine Wissensgeschichte zwischen Leben und Information*. Bielefeld: Transcript, 2009.

Horn, Eva. "Einleitung." In *Schwärme: Kollektive ohne Zentrum: Eine Wissensgeschichte zwischen Leben und Information*, edited by Eva Horn and Lucas Marco Gisi, 7–16. Bielefeld: Transcript, 2009.

Horton, Jessica L. "Drones and Snakes." *Art in America* 105, no. 9 (2017): 104–109.

Houtryve, Tomas van, and Svea Braeunert. "Empathy and the Image under Surveillance Capitalism: Interview with Photographer Tomas van Houtryve." In *Drone Imaginaries: The Remote Power of Vision*, edited by Andreas Immanuel Graae and Kathrin Maurer, 74–88. Manchester: Manchester University Press, 2021.

Howes, David, and Constance Classen. *Ways of Sensing: Understanding the Senses in Society*. London: Routledge, 2013.

Hsu, Hwai-Jung, and Kuan-Ta Chen. "Face Recognition on Drones: Issues and Limitations." In *Proceedings of the First Workshop on Micro Aerial Vehicle Networks, Systems, and Applications for Civilian Use,* edited by Kuan-Ta Chen et al., 39–44. New York: Association for Computing Machinery, 2015.

Hubermann, Georges-Didi. *Das Nachleben der Bilder: Kunstgeschichte und Phantomzeit nach Aby Warburg.* Frankfurt: Suhrkamp, 2010.

Hui, Yuk. "Machine and Ecology." *Angelaki* 25, no. 4 (2020): 54–66.

Huyssen, Andreas. "Fortifying the Heart—Totally: Ernst Jünger's Armored Texts." *New German Critique* 59, no. 1 (1993): 3–23.

Hyde, Ralph. *Panoramania! The Art and Entertainment of the All-Embracing View.* London: Trefoil, 1988.

Ihde, Don. *Bodies in Technology.* Minneapolis: University of Minnesota Press, 2002.

Jablonowski, Maximilian. "Beyond Drone Vision: The Embodied Telepresence of First-Person-View Drone Flight." *Senses and Society* 15, no. 3 (2020): 344–358.

Jablonowski, Maximilian. "Drone It Yourself! On the Decentering of Drone Stories." *Culture Machine* 16 (2015). http://culturemachine.net/drone-culture/drone-ityourself/.

Jablonowski, Maximilian. "Dronie Citizenship?" In *Selfie Citizenship,* edited by Adi Kuntsman, 97–106. London: Palgrave Macmillan, 2017.

Jablonowski, Maximilian. "Ferngesteuertes Feeling: Zur Technogenen Sensualität Unbemannten Fliegens." In *Kulturen der Sinne: Zugänge zur Sensualität der sozialen Welt,* edited by Karl Braun, Claus-Marco Dieterich, Thomas Hengartner, and Bernhard Tschofen, 385–391. Würzburg: Königshausen Neumann, 2017.

Jablonowski, Maximilian. "Der Nomos des Vertikalen: Zur Ortung und Ordnung ziviler Drohnen." In *Ordnungen in Alltag und Gesellschaft: Empirisch-kulturwissenschaftliche Perspektiven,* edited by Stefan Groth and Linda Mülli, 77–92. Würzburg: Königshausen & Neumann, 2019.

Jackman, Anna. "Sensing." Society for Cultural Anthropology, *Theorizing the Contemporary. Fieldsights* (blog), June 27, 2017. https://culanth.org/fieldsights/sensing.

Jacobson, Lindsay. "Be One with the Drone: Elite Athletes Compete Head-to-Head in Drone Racing League." *ABC News,* July 27, 2019. https://abcnews.go.com/Technology/drone-elite-athletes-compete-head-head-drone-racing/story?id=64494623.

Jay, Martin. "Cultural Relativism and the Visual Turn." *Journal of Visual Culture* 1, no. 3 (2002): 267–278.

Jay, Martin. *Downcast Eyes: The Denigration of Vision in Twentieth-Century French Thought.* Berkeley: University of California Press, 1993.

Jay, Martin. "The Scopic Regimes of Modernity." In *Vision and Visuality*, edited by Hal Foster, 3–29. Seattle: Bay Press, 1988.

Jensen, Ole B. "New 'Foucauldian Boomerangs': Drones and Urban Surveillance." *Surveillance & Society* 14, no. 1 (2016): 20–33.

Jensen, Ole B. "Thinking with the Drone: Visual Lessons in Aerial and Volumetric Thinking." *Visual Studies* 35, no. 5 (2020): 417–428.

Johnson, James. "Artificial Intelligence, Drone Swarming and Escalation Risks in Future Warfare." *RUSI Journal* 165, no. 2 (2020): 26–36.

Johnston, John. *The Allure of Machinic Life: Cybernetics, Artificial Life, and the New AI.* Cambridge, MA: MIT Press, 2008.

Johnston, John. "Machinic Vision." *Critical Inquiry* 26 (1999): 27–48.

Jones, Caroline A. "Introduction." In *Sensorium: Embodied Experience, Technology, and Contemporary Art*, edited by Caroline A. Jones, 1–4. Cambridge, MA: MIT Press, 2006.

Jones, Caroline A. "The Mediated Sensorium." In *Sensorium: Embodied Experience, Technology, and Contemporary Art*, edited by Caroline A. Jones, 5–49. Cambridge, MA: MIT Press, 2006.

Jünger, Ernst. *The Glass Bees.* Translated by Louise Bogan and Elizabeth Meyer. New York: Noonday Press, 1960. First published 1957 in German.

Kanz, Christine. *Maternale Moderne: Männliche Gebärphantasien zwischen Kultur und Wissenschaft (1890–1933).* Munich: Fink, 2009.

Kanz, Christine. "Max Beckmann's Revenants and Ernst Jünger's Drones: Vision and Coolness in the Interwar Period." In *Visualizing War*, edited by Anders Engberg-Pedersen and Kathrin Maurer, 43–56. New York: Routledge, 2018.

Kapadia, Ronak K. *Insurgent Aesthetics: Security and the Queer Life of the Forever War.* Durham, NC: Duke University Press, 2019.

Kaplan, Caren. *Aerial Aftermaths: Wartime from Above.* Durham, NC: Duke University Press, 2017.

Kaplan, Caren. "Atmospheric Politics: Protest Drones and the Ambiguity of Airspace." *Digital War* 1 (2020): 50–57.

Kaplan, Caren. "Balloon Geography: The Emotion of Motion in Aerostatic Wartime." In *Aerial Aftermaths: Wartime from Above*, 68–103. Durham: Duke University Press, 2017.

Kaplan, Caren. "The Drone-o-Rama: Troubling the Temporal and Spatial Logics of Distance Warfare." In *Life in the Age of Drone Warfare*, edited by Caren Kaplan and Lisa Parks, 161–177. Durham, NC: Duke University Press, 2017.

Kaplan, Caren. "Eyes in the Skies: *Repellent Fence* and Trans-Indigenous Time-Space at the US-Mexico Border." In *Drone Imaginaries: The Remote Power of Vision*, edited by Andreas Immanuel Graae and Kathrin Maurer, 203–224. Manchester: Manchester University Press, 2021.

Kaplan, Caren. "Mobility and War: The Cosmic View of US Air Power." *Environment and Planning* 38, no. 2 (2006): 395–407.

Kaplan, Caren, and Andrea Miller. "Drones as Atmospheric Policing: From US Border Enforcement to the LAPD." *Public Culture* 31, no. 3 (2019): 419–445.

Kaplan, Caren, and Patricia A. Zimmermann. "Coronavirus Drone Genres: Spectacles of Distance and Melancholia." *Film Quarterly* (April 30, 2020). https://filmquarterly .org/2020/04/30/coronavirus-drone-genres-spectacles-of-distance-and-melancholia/.

Kapp, Ernst. *Elements of a Philosophy of Technology: On the Evolutionary History of Culture*. Minneapolis: University of Minnesota Press, 2018.

Kent, Neil. *The Sami Peoples of the North: A Social and Cultural History*. New York: Hurst & Co., 2018.

Kiesel, Helmuth. *Ernst Jünger: Die Biographie*. Munich: Siedler, 2009.

Kindervater, Katharine Hall. "The Emergence of Lethal Surveillance: Watching and Killing in the History of Drone Technology." *Security Dialogue* 47, no. 3 (2016): 223–238.

Kirkwood, Jeffrey West, and Leif Weatherby. "Operations of Culture: Ernst Kapp's Philosophy of Technology." *Grey Room* (2018): 6–15.

Kirszenbaum, Martha. "Korakrit Arunanondchai by Martha Kirszenbaum." *Bomb* 149 (2019). https://bombmagazine.org/articles/korakrit-arunanondchai/Kirszenbaum.

Kitchin, Rob. "Civil Liberties *or* Public Health, or Civil Liberties *and* Public Health? Using Surveillance Technologies to Tackle the Spread of COVID-19." *Space and Polity* 24, no. 3 (2020): 362–381.

Kitchin, Rob. "Using Digital Technologies to Tackle the Spread of the Coronavirus: Panacea or Folly?" Programmable City Working Paper 44, Maynooth University, Maynooth, April 2020. http://progcity.maynoothuniversity.ie/.

Kittler, Friedrich. *Gramophone, Film, Typewriter*. Stanford: Stanford University Press, 1999.

Klauser, Francisco. "Looking Upwards: Drones and the Social Appropriation of Airspace." In *Game of Drones: Of Unmanned Aerial Vehicles*, edited by Claudia Emmert, Jürgen Bleibler, Ina Neddermeyer, and Dominik Busch, 164–173. Berlin: Neofelis, 2020.

Klauser, Franciso. "Splintering Spheres of Security: Peter Sloterdijk and the Contemporary Fortress City." *Environment and Planning: Society and Space* 28, no. 2 (2010): 326–340.

Klauser, Francisco. *Surveillance and Space*. London: Sage, 2016.

Klauser, Francisco. "Surveillance Farm: Towards a Research Agenda on Big Data Agriculture." *Surveillance & Society* 16, no. 3 (2018): 370–378.

Kolbert, Elizabeth. "Enter the Anthropocene Age of Man." *National Geographic* 219, no. 3 (2011): 60–85.

"Korakrit Arunanondchai: Painting with History in a Room Filled with People with Funny Names 3 at Palais de Tokyo Paris." *Mousse Magazine*, 2015. http://moussemagazine.it/korakrit-arunanondchai-palaisdetokyo.

Koss, Juliet. "On the Limits of Empathy." *Art Bulletin* 88, no. 1 (2006): 139–157.

Kracauer, Siegfried. "The Mass Ornament." In *The Mass Ornament: Weimar Essays*, edited and translated by Thomas Y. Levin, 75–88. Cambridge, MA: Harvard University Press, 1995.

Krämer, Sybille. "Operative Bildlichkeit: Von der Grammatologie zu einer Diagrammatologie? Reflexionen über erkennendes Sehen." In *Logik des Bildlichen: Zur Kritik der ikonischen Vernunft*, edited by Martina Hessler and Dieter Mersch, 94–122. Bielefeld: Transcript, 2009.

Kranzberg, Melvin. "Technology and History: Kranzberg's Laws." *Technology and Culture* 27, no. 3 (1986): 544–560.

Krauss, Rosalind. "Grids." *October* 9 (1979): 51–64.

Kullmann, Karl. "The Drone's Eye: Applications and Implications for Landscape Architecture." *Landscape Research* 43, no. 7 (2018): 906–921.

Kunak, Göksu. "Interview: Hito Steyerl: Zero Probability and the Age of Mass Art Production." *Berlin Art Link*, November 19, 2013. http://www.berlinartlink.com/2013/11/19/interview-hito-steyerl-zero-probability-and-the-age-of-mass-art-production/.

Kurgan, Laura. *Close Up at a Distance: Mapping, Technology, and Politics*. New York: Zone, 2013.

Lacan, Jacques. "The Subversion of the Subject and the Dialectic of Desire in the Freudian Unconscious." In *Écrits: A Selection*, 323–360. New York: Routledge, 1977.

Lamsters, Kristaps, Jānis Karušs, Māris Krievāns, and Jurijs Ješkins. "Application of Unmanned Aerial Vehicles for Glacier Research in the Arctic and Antarctic." *Environment, Technologies, Resources: Proceedings of the International Scientific and Practical Conference* 1 (2019): 131–135.

Latour, Bruno. *Facing Gaia: Eight Lectures on the New Climatic Regime*. Cambridge: Polity, 2017.

Latour, Bruno. "Why Gaia Is Not a God of Totality." *Theory, Culture & Society* 34, nos. 2–3 (2017): 61–81.

Lauzon, Claudette. "Stranger Things: A Techno-Bestiary of Drones in Art and War." In *Drone Imaginaries: The Remote Power of Vision*, edited by Andreas Immanuel Graae and Kathrin Maurer, 180–202. Manchester: Manchester University Press, 2021.

Lechler, Bernd. "Raphaela Vogel, Drohnen: Ungeheuer am Himmel?" Deutschlandfunk, June 6, 2019. https://www.deutschlandfunk.de/raphaela-vogel-bei-game-of -drones-drohnen-ungeheuer-am.807.de.html?dram:article_id=450868.

Lee-Morrison, Lila. *Portraits of Automated Facial Recognition: On Machinic Ways of Seeing the Face*. Bielefeld: Transcript, 2019.

Lethen, Helmut. *Verhaltenslehren der Kälte: Lebensversuche zwischen den Kriegen*. Frankfurt: Suhrkamp, 1994.

Levin, Michael. *Modernity and the Hegemony of Vision*. Berkeley: University of California Press, 1993.

Levinas, Emanuel. *Ethics and Infinity: Conversations with Philippe Nemo*. Translated by Richard A. Cohen. Pittsburgh: Duquesne University Press, 2011.

Liljefors, Max. "Omnivoyance and Blindness." In *War and Algorithm*, edited by Max Liljefors, Gregor Noll, and Daniel Steuer, 127–164. London: Rowman and Littlefield, 2020.

Liljefors, Max, Gregor Noll, and Daniel Steuer, eds. *War and Algorithm*. London: Rowman and Littlefield, 2020.

Lockwood, Jeffrey. *The Infested Mind: Why Humans Fear, Loathe, and Love Insects*. Oxford: Oxford University Press, 2013.

Luke, Timothy. "On the Politics of the Anthropocene." *Telos* 172 (2015): 139–162.

Luo, Shuangling, Haoxiang Xia, Taketoshi Yoshida, and Zhongtuo Wang. "Toward Collective Intelligence of Online Communities: A Primitive Conceptual Model." *Journal of Systems Science and Systems Engineering* 18, no. 2 (2009): 203–221.

Lyon. David, ed. *Surveillance as Social Sorting: Privacy, Risk, and Digital Discrimination*. New York: Routledge, 2003.

Lyon, David. "Introduction." In *Surveillance as Social Sorting*, edited by David Lyon, 1–9. New York: Routledge, 2003.

Lyon, David. "Surveillance and Social Sorting: Computer Codes and Social Sorting." In *Surveillance as Social Sorting*, edited by David Lyon, 13–20. New York: Routledge, 2003.

Lyotard, Jean-François. *Lessons on the Analytic of the Sublime: Kant's Critique of Judgment*. Translated by Elizabeth Rottenberg. Stanford: Stanford University Press, 1994.

Magnet, Shoshana. *When Biometrics Fail: Gender, Race, and the Technology of Identity*. Durham, NC: Duke University Press, 2011.

Majetschak, Stefan. "Sichtvermerke: Über den Unterschied zwischen Kunst und Gebrauchsbildern." In *Bild-Zeichen: Perspektiven einer Wissenschaft vom Bild*, edited by Stefan Majetschak, 97–121. Munich: Fink, 2005.

Manderson, Desmond. "Chronotopes in the Scopic Regime of Sovereignty." *Visual Studies* 32, no. 2 (2017): 167–177.

Mann, Steve, Jason Nolan, and Barry Wellman. "Sousveillance: Inventing and Using Wearable Computing Devices for Data Collection in Surveillance Environments." *Surveillance & Society* 1, no. 3 (2003): 331–355.

Manning, Erin. *Politics of Touch: Sense, Movement, Sovereignty*. Minneapolis: University of Minnesota Press, 2007.

Manovich, Lev. *The Language of New Media*. Cambridge, MA: MIT Press, 2005.

Massumi, Brian. *Ontopower: War, Powers, and the State of Perception*. Durham, NC: Duke University Press, 2015.

Maurer, Kathrin. "Adalbert Stifter's Poetics of Clouds and Nineteenth-Century Meteorology." *Oxford German Studies* 45, no. 4 (2016): 421–433.

Maurer, Kathrin. "Ballooning as a Technology of Seeing in Jean Paul's *Des Luftschiffers Giannozzo Seebuch* (1801)." In *Before Photography*, edited by Kirsten Belgum, Vance Byrd, and John D. Benjamin, 17–38. Berlin: Walter de Gruyter, 2021.

Maurer, Kathrin. "Det Farlige Øjeblik: Ernst Jüngers Fotobøger og hans Teori om Fotografiet." In *Soldat, Arbejder, Anark: Ernst Jüngers Forfatterskab*, edited by Adam Paulsen, 155–176. Copenhagen: Museum Tusculanum, 2017.

Maurer, Kathrin. "Drones as Big Data Archives: Mimesis and Counter-Archiving in Contemporary Art on Military Drones." In *(W)archives*, edited by Daniela Agostinho, Solveig Gade, Nanna Bonde Thylstrup, and Kristin Veel, 119–141. Berlin: Sternberg Press, 2021.

Maurer, Kathrin. "Flattened Vision: Nineteenth-Century Hot Air Balloons as Early Drones." In *Drone Imaginaries: The Remote Power of Vision*, edited by Andreas Immanuel Graae and Kathrin Maurer, 19–38. Manchester: Manchester University Press, 2021.

Maurer, Kathrin. "The Paradox of Total Immersion: Watching War in Nineteenth-Century Panoramas." In *Visualizing War: Images, Emotions, Communities*, edited by Anders Engberg-Pedersen and Kathrin Maurer, 78–94. New York: Routledge, 2018.

Maurer, Kathrin. "Translating Catastrophes: Yoko Tawada's Poetic Responses to the 2011 Tōhoku Earthquake, the Tsunami, and Fukushima." *New German Critique* 43, no. 1 (2016): 171–194.

Maurer, Kathrin. "Visual Power: The Scopic Regime of Military Drones." *War, Media, Conflict* 10, no. 2 (2017): 141–151.

Maurer, Kathrin. *Visualizing the Past: The Power of the Image in German Historicism.* Berlin: Walter de Gruyter, 2013.

Maxwell, Richard, and Toby Miller. *Greening the Media.* Oxford: Oxford University Press, 2012.

McCosker, Anthony. "Drone Media: Unruly Systems, Radical Empiricism, and Camera Consciousness." *Culture Machine* 16 (2015). https://culturemachine.net/drone-culture/drone-media/.

McFarland, Matt. "Slaughterbots Film Shows Potential Horrors of Killer Drones." *CNNMoney*, November 14, 2017. https://money.cnn.com/2017/11/14/technology/autonomous-weapons-ban-ai/index.html.

McLean, Heather. "In Praise of Chaotic Research Pathways: A Feminist Response to Planetary Urbanization." *Environment and Planning: Society and Space* 36, no. 3 (2018): 547–555.

McNeil, Joanne, and Ingrid Burrington. "Dronism." *Dissent* 61, no. 2 (2014): 57–60.

Meschiari, Matteo. "Roots of the Savage Mind: Apophenia and Imagination as Cognitive Process." *Quaderni di Semantica* 2, no. 1 (2009): 1–39.

Metz, Christian. *Film Language: A Semiotics of the Cinema.* Translated by Michael Taylor. Chicago: University of Chicago Press, 1991. First published 1971 in French.

Mieszkowski, Jan. "The Drone of Data." In *Drone Imaginaries: The Remote Power of Vision*, edited by Andreas Immanuel Graae and Kathrin Maurer, 55–73. Manchester: Manchester University Press, 2021.

Millar, Heather. "Racing Drones." *Air & Space Magazine*, November 2016. https://www.airspacemag.com/flight-today/racing-drones-180960969/.

Milligan, Brett. "Making Terrains: Surveying, Drones and Media Ecology." *Journal of Landscape Architecture* 14, no. 2 (2019): 20–35.

Minsky, Marvin. "Telepresence." *Omni Magazine*, June 1980. https://web.media.mit.edu/~minsky/papers/Telepresence.html.

Mirzoeff, Nicholas. "The Right to Look." *Critical Inquiry* 37, no. 3 (2011): 473–496.

Mirzoeff, Nicholas. "Visualizing the Anthropocene." *Public Culture* 26, no. 2 (2014): 213–232.

Misawa, Kana, and Jun Rekimoto. "ChameleonMask: Embodied Physical and Social Telepresence Using Human Surrogates." In *Proceedings of the 33rd Annual ACM Conference: Extended Abstracts on Human Factors in Computing Systems*, edited by Bo Begole and Jinwoo Kim, 401–411. New York: Association for Computing Machinery, 2015.

Mishara, Aaron L. "Klaus Conrad (1905–1961): Delusional Mood, Psychosis, and Beginning Schizophrenia." *Schizophrenia Bulletin* 36, no. 1 (2010): 9–13.

Mitchell, W. J. T. *What Do Pictures Want? The Lives and Loves of Images*. Chicago: University of Chicago Press, 2005.

Miyoshi, Masao. "Turn to the Planet: Literature, Diversity, and Totality." *Comparative Literature* 53, no. 4 (2001): 283–297.

Moore, Tina, and Dean Balsamini. "NYPD Seizes Drone Documenting Mass Hart Island Burials amid Coronavirus." *New York Post*, April 18, 2020. https://nypost.com/2020/04/18/nypd-seizes-drone-documenting-mass-hart-island-burials-amid-coronavirus.

Moraru, Christian. *Reading for the Planet: Toward a Geomethodology*. Ann Arbor: University of Michigan Press, 2015.

Morris, Kate. *Shifting Grounds: Landscape in Contemporary Native American Art*. Seattle: Washington University Press, 2019.

Morton, Timothy. *The Ecological Thought*. Boston, MA: Harvard University Press, 2010.

Morton, Timothy. *Ecology without Nature: Rethinking Environmental Aesthetics*. Cambridge, MA: Harvard University Press, 2007.

Morton, Timothy. *Hyperobjects: Philosophy and Ecology after the End of the World*. Minneapolis: University of Minnesota Press, 2013.

Næss, Arne, and Bob Jickling. "Deep Ecology and Education: A Conversation with Arne Næss." *Canadian Journal of Environmental Education* 5, no. 1 (2000): 48–62.

Nancy, Jean-Luc. "The Confronted Community." Translated by Amanda Macdonald. *Postcolonial Studies* 6, no. 1 (2003): 23–36.

Nancy, Jean-Luc. *The Inoperative Community*. Edited by Peter Connor, Lisa Garbus, Michael Holland, and Simona Sawhney. Translated by Peter Connor. Minneapolis: University of Minnesota Press, 1991.

Neaman, Elliot. *A Dubious Past: Ernst Jünger and the Politics of Literature after Nazism*. Berkeley: University of California Press, 1999.

Ngai, Sianne. *Ugly Feelings*. Cambridge, MA: Harvard University Press, 2009.

Nochlin, Linda. *The Politics of Vision: Essays on Nineteenth-Century Art and Society*. Boulder: Westview Press, 1989.

Nora, Pierre, and Lawrence D. Kritzman, eds. *Realms of Memory: Rethinking the French Past*. Vols. 1–3. Translated by Arthur Goldhammer. New York: Columbia University Press, 1996.

Noys, Benjamin. "Drone Metaphysics." *Culture Machine* 16 (2015): 1–22.

Nye, David E. *American Technological Sublime*. Cambridge, MA: MIT Press, 1996.

Oettermann, Stephan. *The Panorama: History of a Mass Medium*. Translated by Deborah Lucas-Schneider. New York: Zone, 1997.

Olsen, Hanna Brooks. "Dronie Like a Pro: How to Master the Drone Selfie." CreativeLive , July 21, 2014. https://www.creativelive.com/blog/dronie-tips.

Paglen, Trevor. *I Could Tell You But Then You Would Have to Be Destroyed by Me: Emblems from the Pentagon's Dark World*. Brooklyn: Melville House, 2006.

Paglen, Trevor. "Invisible Images: Your Pictures Are Looking at You." *Architectural Design* 89, no. 1 (2019): 22–27.

Parikka, Jussi. *A Geology of Media*. Minneapolis: University of Minnesota Press, 2015.

Parikka, Jussi. *Insect Media: An Archaeology of Animals and Technology*. Minneapolis: University of Minnesota Press, 2010.

Parikka, Jussi. *Medianatures*. London: Open Humanities Press, 2013.

Parikka, Jussi. "Politics of Swarms: Translations between Entomology and Biopolitics." *Parallax* 14, no. 3 (2008): 112–124.

Parks, Lisa. *Cultures in Orbit: Satellites and the Televisual*. Durham: Duke University Press, 2005.

Parks, Lisa. "Drones, Infrared Imagery, and Body Heat." *International Journal of Communication* 8 (2014): 2518–2521.

Parks, Lisa. "Drones, Vertical Mediation, and the Targeted Class." *Feminist Studies* 42, no. 1 (2016): 227–235.

Parks, Lisa. "Vertical Mediation and the U.S. Drone War in the Horn of Africa." In *Life in the Age of Drone Warfare*, edited by Caren Kaplan and Lisa Parks, 134–158. Durham, NC: Duke University Press, 2017.

Petersen, Rikke Munck. "The Dispatched Drone and Affective Distance in Fieldwork." *The Senses and Society* 15, no. 3 (2020): 311–328.

Pizer, John. "Planetary Poetics: World Literature, Goethe, Novalis, and Yoko Tawada's Translational Writing." In *The Planetary Turn: Relationality and Geoaesthetics in the Twenty-First Century*, edited by Amy J. Elias and Christian Moraru, 3–24. Evanston: Northwestern University Press, 2015.

Ploeg, Irma van der. "Biometrics and the Body as Information." In *Surveillance as Social Sorting: Privacy, Risk, and Digital Discrimination*, edited by David Lyon, 57–73. New York: Routledge, 2003.

Puig de la Bellacasa, Maria. *Matters of Care: Speculative Ethics in More Than Human Worlds*. Minneapolis: Minnesota University Press, 2017.

Rancière, Jacques. *Dissensus: On Politics and Aesthetics*. Edited and translated by Steven Corcoran. London: Continuum, 2010.

Rancière, Jacques. *The Politics of Aesthetics*. Translated by Gabriel Rockhill. London: Bloomsbury, 2010.

Raunig, Gerald. *Dividuum: Machinic Capitalism and Molecular Revolution*. Translated by Aileen Derieg. South Pasadena: Semiotext(e), 2016.

Rebentisch, Juliane. *Ästhetik der Installation*. Frankfurt: Suhrkamp, 2003.

Reid, Paul. *Biometrics for Network Security*. Upper Saddle River: Prentice Hall, 2004.

Rhee, Jennifer. "Adam Harvey's 'Anti-Drone' Wear in Three Sites of Opacity." *Camera Obscura: Feminism, Culture, and Media Studies* 31, no. 2 (2016): 175–185.

Richardson, Michael. "Drone Cultures: Encounters with Everyday Militarisms." *Continuum* 34, no. 6 (2020): 858–869.

Richardson, Michael. "Pandemic Drones." The Conversation, March 31, 2020. https://theconversation.com/pandemic-drones-useful-for-enforcing-social-distancing-or-for-creating-a-police-state-134667.

Richardson, Niall, and Adam Locks. *Body Studies: The Basics*. New York: Routledge, 2014.

Ritter, Joachim, Karlfried Gründer, and Gottfried Gabriel, eds. *Historisches Wörterbuch der Philosophie*. Vol. 1. Basel: Schwabe, 2005.

Robbins, Christopher. "NYPD Seizes Drone of Photojournalist Documenting Mass Burials on Hart Island." Gothamist, April 20, 2020. https://gothamist.com/news/nypd-seizes-drone-photojournalist-documenting-mass-burials-hart-island.

Robinson, Kim Stanley. *The Ministry for the Future*. New York: Orbit Books, 2020.

Roßler, Gustav. "Kleine Galerie neuer Dingbegriffe: Hybriden, Quasi-Objekte, Grenzobjekte, epistemische Dinge." In *Bruno Latours Kollektive: Kontroversen zur Entgrenzung des Sozialen*, edited by Georg Kneer, Markus Schroer, and Erhard Schüttpelz, 76–107. Frankfurt: Suhrkamp, 2008.

Russell, Stuart, Anthony Aguirre, Ariel Conn, and Max Tegmark. "Why You Should Fear Slaughterbots." *IEEE Spectrum*, January 23, 2018. https://spectrum.ieee.org/automaton/robotics/artificial-intelligence/why-you-should-fear-slaughterbots-a-response.

Saif, Atef Abu. *The Drone Eats with Me: Diaries from a City under Fire*. Manchester: Comma Press, 2015.

Salter, Michael. "Toys for the Boys? Drones, Pleasure and Popular Culture in the Militarisation of Policing." *Critical Criminology* 22, no. 2 (2014): 163–177.

Sands, Daniele. "Gaia Politics, Critique, and the Planetary Imaginary." *Substance* 49, no. 3 (2020): 104–121.

Sandvik, Kristin Bergtora. "African Drone Stories." *Behemoth: A Journal on Civilisation* 8, no. 2 (2015): 73–96.

Sandvik, Kristin Bergtora, and Maria Gabrielsen Jumbert, eds. *The Good Drone*. New York: Routledge, 2016.

Sandvik, Kristin Bergtora, and Maria Gabrielsen Jumbert. "Humanitarian Drones: An Inventory." *Revue Internationale et Stratégique* 98, no. 2 (2015): 139–146.

Sandvik, Kristin Bergtora, Katja Lindskov Jacobsen, and Sean Martin McDonald. "Do No Harm: A Taxonomy of the Challenges of Humanitarian Experimentation." *International Review of the Red Cross* 99, no. 904 (2017): 319–344.

Sarlos, Garbor. "Hated in the Nation: A Phantasma of the Post-Climate World." In *Reading Black Mirror: Insights into Technology and the Post-Media Condition*, edited by German A. Duarte and Justin Micheal Battin, 309–324. Bielefeld: Transcript, 2021.

Schaefer, John D. *Sensus Communis: Vico, Rhetoric, and the Limits of Relativism*. Durham, NC: Duke University Press, 1990.

Scharre, Paul. *Army of None: Autonomous Weapons and the Future of War*. New York: Norton, 2018.

Scharre, Paul. "How Swarming Will Change Warfare." *Bulletin of the Atomic Scientists* 74, no. 6 (2018): 385–389.

Scharre, Paul. "Why You Should Not Fear Slaughterbots." *IEEE Spectrum*, December 22, 2017. https://spectrum.ieee.org/automaton/robotics/military-robots/why-you -shouldnt-fear-slaughterbots.

Schinkel, Eckhard. "Der Ballon in der Literatur." In *Leichter als Luft: Zur Geschichte der Ballonfahrt*, edited by Bernard Korzus and Gisela Noehles, 200–236. Münster: Westfälisches Landesmuseum für Kunst und Kulturgeschichte, 1978.

Schiølin, Kasper. "Revolutionary Dreams: Future Essentialism and the Sociotechnical Imaginary of the Fourth Industrial Revolution in Denmark." *Social Studies of Science* 50, no. 4 (2020): 542–566.

Schivelbusch, Wolfgang. *The Railway Journey*. New York: Urizen, 1980.

Schmarsow, August. "The Essence of Architectonic Creation." In *Empathy, Form, and Space*, edited by Harry Francis Mallgrave and Eleftherios Ikonomou, 281–297. Santa Monica: Getty Center for the History of Art and the Humanities, 1994.

Schmidt-Burckhardt, Astrit. "The All-Seer: God's Eye as Proto-Surveillance." In *CRTL [SPACE]: Rhetorics of Surveillance from Bentham to Big Brother*, edited by Thomas Y. Levin, Ursula Frohne, and Peter Weibel, 16–31. Cambridge, MA: MIT Press, 2002.

Schnepf, J. D. "Flood from Above: Disaster Mediation and Drone Humanitarianism." *Media + Environment* 2, no. 1 (2020): 13466.

Schnepf, J. D. "Unsettling Aerial Surveillance: Surveillance Studies after Standing Rock." *Surveillance & Society* 17, no. 5 (2019): 747–751.

Schoofs, Hilde, Stephanie Delalieux, Tom Deckers, and Dany Bylemans. "Fire Blight Monitoring in Pear Orchards by Unmanned Airborne Vehicles (UAV) Systems Carrying Spectral Sensors." *Agronomy* 10, no. 5 (2020): 1–12.

Segal, Howard P. *Technological Utopianism in American Culture*. Syracuse: Syracuse University Press, 2005.

Shackle, Samira. "The Mystery of the Gatwick Drone." *The Guardian*, December 1, 2020. https://www.theguardian.com/uk-news/2020/dec/01/the-mystery-of-the-gatwick-drone.

Shane, Scott, and Daisuke Wakabayashi. "The Business of War: Google Employees Protest Work for the Pentagon." *New York Times*, April 4, 2018. https://www.nytimes.com/2018/04/04/technology/google-letter-ceo-pentagon-project.html.

Shaw, Ian. *Predator Empire: Drone Warfare and Full Spectrum Dominance*. Minneapolis: University of Minnesota Press, 2016.

Siegert, Bernd. *Cultural Techniques: Grids, Filters, Doors, and Other Articulations of the Real*. Translated by Geoffrey Winthrop-Young. New York: Fordham University Press, 2015.

Siegert, Bernd. "(Not) in Place: The Grid, or, the Cultural Techniques of Ruling Spaces." In *Cultural Techniques: Grids, Filters, Doors, and Other Articulations of the Real*, 97–120. New York: Fordham University Press, 2015.

Simondon, Gilbert. *On the Mode of Existence of Technical Objects*. Translated by Cecile Malaspina and John Rogove. Minneapolis: University of Minnesota Press, 2017.

Simonite, Tom. "Facebook Creates Software That Matches Faces Almost as Well as You Do." *Technology Review*, March 17, 2014. https://www.technologyreview.com/2014/03/17/13822/facebook-creates-software-that-matches-faces-almost-as-well-as-you-do.

Sloterdijk, Peter. *Terror from the Air*. Translated by Amy Patton and Steve Corcoran. Los Angeles: Semiotext(e), 2009.

Smith, Terry. "Comparing Contemporary Arts; or, Figuring Planetarity." In *The Planetary Turn: Relationality and Geoaesthetics in the Twenty-First Century*, edited by Amy J. Elias and Christian Moraru, 175–192. Evanston: Northwestern University Press, 2015.

Somaini, Antonio. "Machine Vision in Pandemic Times." In *Pandemic Media: Preliminary Notes Toward an Inventory*, edited by Philipp Dominik Keidl, Laliv Melamed, Vinzenz Hediger, and Antonio Somaini, 147–156. Lüneburg: Meson Press, 2020.

Somaini, Antonio. "On the Scopic Regime." *Leitmotiv* 5 (2005–2006): 25–38.

Soroye, Peter, Tim Newbold, and Jeremy Kerr. "Climate Change Contributes to Widespread Declines among Bumble Bees across Continents." *Science* 367, no. 6478 (2020): 685–688.

South China Morning Post. *Shanghai's Drone Show Welcoming 2020 Reportedly Never Happened on New Year's Eve*. YouTube, 2:25, uploaded January 30, 2020. https://www.youtube.com/watch?v=F_DkUXiLczE.

Spivak, Gayatri Chakravorty. *Death of a Discipline*. New York: Columbia University Press, 2003.

Stahl, Roger. "What the Drone Saw: The Cultural Optics of the Unmanned War." *Australian Journal of International Affairs* 67, no. 5 (2013): 659–674.

Steinmetz, George. "Drones Are Changing How We See the World." *Time*, May 31, 2018. https://time.com/longform/drones-career.

Steinmetz, George, and Andrew Revkin. *The Human Planet: Earth at the Dawn of the Anthropocene*. New York: Abrams, 2020.

Steyerl, Hito. "In Defense of the Poor Image." *e-flux journal* 11 (2009). http://www.e-flux.com/journal/view/94.

Steyerl, Hito. "In Free Fall: A Thought Experiment on Vertical Perspective." *e-flux journal* 24 (2011): 1–11.

Steyerl, Hito. "In Free Fall: A Thought Experiment on Vertical Perspective." In *To See without Being Seen: Contemporary Art and Drone Warfare*, edited by Svea Braeunert and Meredith Malone, 71–81. Chicago: University of Chicago Press, 2016.

Steyerl, Hito. "A Sea of Data: Apophenia and Pattern (Mis-)recognition." *e-flux journal* 72 (2016): 1–14.

Steyerl, Hito, and Laura Poitras. "Techniques of the Observer: Hito Steyerl and Laura Poitras in Conversation." *Artforum International* 53, no. 9 (2015): 306–317.

Stiegler, Bernard. *The Neganthropocene*. Edited and translated by Daniel Ross. London: Open Humanities Press, 2018.

Strawser, Bradley J. "Moral Predators: The Duty to Employ Uninhabited Aerial Vehicles." *Journal of Military Ethics* 9, no. 4 (2015): 342–368.

Stubblefield, Thomas. *Drone Art: The Everywhere War as Medium*. Oakland: University of California Press, 2020.

Stubblefield, Thomas. "In Pursuit of Other Networks." In *Life in the Age of Drone Warfare*, edited by Caren Kaplan and Lisa Parks, 195–219. Durham, NC: Duke University Press, 2017.

Suchman, Lucy. *Human-Machine Reconfigurations: Plans and Situated Actions*. New York: Cambridge University Press, 2007.

Suchman, Lucy, and Jutta Weber. "Human-Machine Autonomies." In *Autonomous Weapon Systems: Law, Ethics, Policy*, edited by Bhuta Nehal, Susanne Beck, Robin Geiβ, 75–102. New York: Cambridge University Press, 2016.

Tahir, Anam, Jari Böling, Mohammad-Hashem Haghbayan, Hannu T. Toivonen, and Juha Plosila. "Swarms of Unmanned Aerial Vehicles: A Survey." *Journal of Industrial Information Integration* 16 (2019): 100106.

Taylor, Charles. *Modern Social Imaginaries*. Durham, NC: Duke University Press, 2003.

Thacker, Eugene. "Networks, Swarms, and Multitudes." *Ctheory.net*, May 18, 2004. https://journals.uvic.ca/index.php/ctheory/article/view/14541/5388.

Tidman, Zoe. "Coronavirus: Italian Mayor Plans on Using Drones to Send People Back Home during Lockdown." *Euronews*, March 26, 2020. https://www.euronews.com/2020/03/26/watch-italian-mayor-uses-drone-to-scream-at-locals-to-stay-indoors-amid-coronavirus-crisis.

Tönnies, Ferdinand. "Gemeinschaft und Gesellschaft (Erstausgabe 1887)." In *Ferdinand Tönnies Gesamtausgabe*, vol. 2, edited by Bettina Clausen and Dieter Haselbach, 3–34. Berlin: Walter de Gruyter, 2019.

Trecka, Mark. "The Implication of a Fence: Part Three: The Sovereignty of Context." *Beacon Broadside* (blog), July 6, 2016. https://www.beaconbroadside.com/broadside/2016/07/the-implication-of-a-fence-part-three-the-sovereignty-of-context.html.

Tuck, Sarah. "Drone Alliances." In *Fragmentation of the Photographic Image in the Digital Age*, edited by Daniel Rubenstein, 73–80. New York: Routledge, 2020.

Tuck, Sarah, ed. *Drone Vision: Warfare, Surveillance, Protest*. Gothenburg: Art and Theory Publishing Hasselblad Foundation, 2022.

Tuck, Sarah. "Drone Vision and Protest." *Photographies* 11, nos. 2–3 (2018): 169–175.

Valaouris, Michael. *Das Feld hat Augen: Bilder des überwachenden Blicks*. Berlin: Deutscher Kunstverlag Kunstbibliothek Staatliche Museen zu Berlin, 2017.

Vandinther, Jackie. "Drone Video Shows Inmates Digging Mass Burial Graves on New York's Hart Island." *CTV News*, April 8, 2020. https://www.ctvnews.ca/world/drone-video-shows-inmates-digging-mass-burial-graves-on-new-york-s-hart-island-1.4888134.

Varley-Winter, Olivia. "The Overlooked Governance Issues Raised by Facial Recognition." *Biometric Technology Today* 5 (2020): 5–8.

Vavarella, Emilio. "Interview with the Drone: Experimenting with Post-Anthropocentric Art Practice." *Digital Creativity* 27, no. 1 (2016): 71–81.

Veys, Charles, James Hibbert, Phillip Davis, and Bruce Grieve. "An Ultra-Low-Cost Active Multispectral Crop Diagnostics Device." In *2017 IEEE Sensors (Proceedings)*, edited by Krikor Oznayan, 1005–1008. New York: Institute of Electrical and Electronics Engineers, 2017.

Villazana, Libia. "Transnational Virtual Mobility as a Reification of Deployment of Power: Exploring Transnational Processes in the Film *Sleep Dealer*." *Transnational Cinemas* 4, no. 2 (2013): 217–230.

Virgillito, Maria Enrica. "Rise of the Robots: Technology and the Threat of a Jobless Future." *Labor History* 58, no. 2 (2017): 240–242.

Virilio, Paul. *War and Cinema: The Logistics of Perception*. Translated by Patrick Camiller. London: Verso, 1989.

Vischer, Robert. "On the Optical Sense of Form: A Contribution to Aesthetics." In *Empathy, Form, and Space: Problems in German Aesthetics, 1973–1893*, edited by Harry Francis Mallgrave and Eleftherios Ikonomou, 89–123. Santa Monica: Getty Center for the History of Art and the Humanities, 1994.

Warburg, Aby. *Der Bilderatlas Mnemosyne*. Berlin: Akademie, 2008.

Wayman, James L. "The Scientific Development of Biometrics over the Last 40 Years." In *The History of Information Security*, edited by Karl De Leeuw and Jan Bergstra, 263–274. Amsterdam: Elsevier Science, 2007.

Weber, Jutta. "Artificial Intelligence and the Socio-Technical Imaginary: On Skynet, Self-Healing Swarms and *Slaughterbots*." In *Drone Imaginaries: The Remote Power of Vision*, edited by Andreas Immanuel Graae and Kathrin Maurer, 167–179. Manchester: Manchester University Press, 2021.

Weizman, Eyal. *Forensic Architecture: Violence at the Threshold of Detectability*. New York: Zone, 2019.

Werber, Niels. "Ants and Aliens: An Episode in the History of Entomological and Sociological Construction of Knowledge." *Berichte zur Wissenschaftsgeschichte* 34, no. 3 (2011): 242–262.

Werber, Niels. "Jüngers Bienen." *Zeitschrift für deutsche Philologie* 130 (2011): 245–260.

Westport Police Department. "Westport Police Department Testing New Drone Technology 'Flatten the Curve' Pilot Program." Press Release, April 2020.

Wich, Serge, Lorna Scott, and Lian Pin Koh. "Wings for Wildlife: The Use of Conservation Drones, Challenges and Opportunities." In *The Good Drone*, edited by Kristin Bergtora Sandvik and Maria Gabrielsen Jumbert, 163–177. New York: Routledge, 2016.

Wiener, Norbert. *Cybernetics or Control and Communication in the Animal and the Machine*. Cambridge, MA: MIT Press, 2019. First published 1948.

Wilcox, Lauren. "Drones, Swarms and Becoming-Insect: Feminist Utopias and Posthuman Politics." *Feminist Review* 116, no. 1 (2017): 25–45.

Wilcox, Lauren. "The Gender Politics of the Drone." In *Drone Imaginaries: The Remote Power of Vision*, edited by Andreas Immanuel Graae and Kathrin Maurer, 110–129. Manchester: Manchester University Press, 2021.

Williams, Alex. "The Drones Were Ready for This Moment." *New York Times*, May 23, 2020. https://www.nytimes.com/2020/05/23/style/drones-coronavirus.html.

Williams, Geoffrey R., and David R. Tarpy, "Colony Collapse Disorder in Context." *Bioessays* 32, no. 10 (2010): 845.

Witte, Karsten, Barbara Correll, and Jack Zipes. "Introduction to Siegfried Kracauer's 'The Mass Ornament.'" *New German Critique* 5, no. 1 (1975): 59–66.

Wittgenstein, Ludwig. *Philosophische Untersuchungen*. Frankfurt: Suhrkamp, 2001.

Wölfflin, Heinrich. "Prolegomena to a Psychology of Architecture." In *Empathy, Form, and Space*, edited by Harry Francis Mallgrave and Eleftherios Ikonomou, 149–187. Santa Monica: Getty Center for the History of Art and the Humanities, 1994.

Young, Liam. *Machine Landscapes: Architectures of the Post Anthropocene*. Hoboken: John Wiley & Sons, 2019.

Zimmermann, Yvonne. "Videoconferencing and the Uncanny Encounter with One-self: Self-Reflexivity as Self-Monitoring 2.0." In *Pandemic Media: Preliminary Notes Toward an Inventory*, edited by Philipp Dominik Keidl, Laliv Melamed, Vinzenz Hediger, and Antonio Somaini, 99–103. Lüneburg: Meson Press, 2020.

Zittel, Claus. *Theatrum Philosophicum: Descartes und die Rolle ästhetischer Formen in der Wissenschaft*. Berlin: Akademie, 2009.

Zuboff, Shosana. *The Age of Surveillance Capitalism: The Fight for a Human Future at the New Frontier of Power*. London: Profile, 2019.

Zuev, Dennis, and Gary Bratchford. "The Citizen Drone: Protest, Sousveillance and Droneviewing." *Visual Studies* 35, no. 5 (2020): 442–456.

Zylinska, Joanna. *Nonhuman Photography*. Cambridge, MA: MIT Press, 2017.

Zylinska, Joanna. "Photography after Extinction." In *After Extinction*, edited by Richard Grusin, 51–70. Minneapolis: University of Minnesota Press, 2018.

Index